# SAXIFRAGES

## THE GENUS SAXIFRAGA IN THE WILD AND IN CULTIVATION

by WINTON HARDING

Joint Hon. Editors:
R. C. Elliott, F.L.S.
C. Grey-Wilson, B.Sc.(Hort.)

©

First Published Autumn 1970 by
The Alpine Garden Society
Lye End Link, St. John's, Woking, Surrey

First Reprint 1976

Second Reprint 1981

ISBN 0 900048 36 0

# THE AUTHOR

William Francis Winton Harding, O.B.E., F.L.S., B.Sc.(Hort.) was born in 1917: he is married, with a nicely balanced family of two boys and two girls. Educated at Aldenham and Reading University he became interested in plants at an early age, working at several nurseries before the war in order to acquire practical experience.

At the close of hostilities (he held a commission in the Royal Corps. of Signals, serving in Egypt and Palestine) he joined the Commonwealth War Graves Commission and for ten years worked in France and Belgium on the planning and planting of the War Cemeteries of northern Europe. In 1956 he became the Commission's Chief Horticultural Officer, and has, ever since, controlled their horticultural work throughout the world. He was appointed an O.B.E. for these services in 1960. He is, needless to say, a widely-travelled man and has, in addition, spent many holidays in the European Mountains studying the alpine plants which are his hobby.

The Alpine Garden Society is most fortunate that Winton Harding should have made a special study of saxifrages, and his considerable knowledge of botany has enabled him to present us with an authoritative and yet eminently readable guide to the genus.

# FOREWORD

The following chapters dealing with the very varied Sections of this complex and fascinating genus have been written over a period of more than two years and have been printed as a series in the *Bulletin* of the Alpine Garden Society. I have endeavoured to tread a middle path between over-simplification and over-complication but am aware that, in so doing, I may have ended up by pleasing neither the gardener nor the botanist. That, however, is a risk I have had to take. It has been my aim that this *Guide* should serve as a useful book of reference for all gardeners who specialize in the cultivation of alpines and that they should be able to find in it the necessary information to grow any saxifrage species currently available. In spite of this aim, I am sure that there is much in these pages which is capable of useful extension or modification. I shall therefore be most grateful to hear from any reader who can, from his or her personal experience, give additional useful information on the growing of any of the species mentioned or omitted.

I wish to take this opportunity of thanking most warmly the Honorary Editor, Roy Elliott and his Assistant Editor, Christopher Grey-Wilson for the very painstaking care they have taken in editing my text and compiling an index.

I have been most fortunate in having the assistance of D. B. Lowe who has specially executed a series of delightful drawings to illustrate the text and I now express to him my sincere appreciation of his whole-hearted co-operation.

Last but not least I thank all those, including Roy Elliott, Miss D. Holford, Miss M. G. Hodgman and D. F. Merrett, who have allowed their photographs to be used in this publication. Happy the author indeed who can call upon such splended prize-winning studies to bring his words to life.

Autumn, 1970                                                    WINTON HARDING

## FOREWORD TO REVISED EDITION

Minor additions and corrections have been made to the text but to avoid confusion and inconvenience to horticulturists I have not altered the names of the Sections as this would eliminate Kabschia as a Section name. The currently recognised valid botanical names are however Section 3 Gymnopera D Don, Section 7 Saxifraga, Section 11 Aizoonia Tausch, Section 12 Porophyllum Gaudin, Section 14 now merged with Section 12, Section 16 Discogyne Sternb (Zahlbrucknera Reichenb) new Section added.

Summer, 1976                                                    WINTON HARDING

# I. INTRODUCTION

WHEN I AGREED TO A REQUEST BY YOUR EDITOR to undertake a series of articles on the genus *Saxifraga*, section by section, I must have been in either a very rash or a very compliant mood for it is a task that would tax the knowledge of a life-time expert. I must therefore crave the reader's indulgence if I fall short of doing this formidable subject the justice which is its due. My excuse must be that for some years now the genus has engaged my fascinated attention and I have taken the opportunity of growing any member of it which came my way. This however has still left many species introduced to horticulture which I have not grown and here I must thank many colleagues for information freely given. It is my earnest hope that enthusiasm and interest may in some measure compensate for the inevitable gaps in my knowledge.

I must admit that, in undertaking this series of articles, I am also hoping to revive interest in a genus where I have felt that it has recently been tending to flag. In Farrer's day the genus *Saxifraga* was, in company with *Campanula* and *Dianthus*, rated as one of the 'big three' making up the essential backbone planting of the alpine garden. Since those days alpine gardeners have accepted the challenge afforded by many other exciting genera and it is good that they should have done so. There is however a place for the old as well as the new, and it is neither right nor healthy for our Society that interest should be concentrated too exclusively on the esoteric. It is probable that the very good temper of many members of this genus has led us to undervalue the interest of the whole. It should not be forgotten that the genus includes, as well as many 'everyday rock plants', many beautiful, difficult and lesser known species to challenge the skill of the plantsman. The day will also assuredly dawn when the many as yet unintroduced Far Eastern species will be introduced or re-introduced for us and it is essential that some of them should then become more than birds of brief passage.

The first problem which has engaged my attention has been the method of treatment to be adopted in these articles. How deeply botanical should this be? On consideration I have felt that it would be an insult to the intelligence of readers of the *Bulletin* were I not to provide the essential taxonomic framework without which it is most difficult to produce order from the chaos of the many hundreds of species which make up this genus. In so doing, however, I have felt that I must never lose sight of the fact that the articles are written for the benefit of intelligent gardeners rather than botanists. Any botanist who therefore reads these notes must bear with me in any omissions or simplifications. The most important omission has been the virtual exclusion of all species which have not been satisfactorily introduced into cultivation and which are chiefly known from herbarium material and field notes.

The classification is based on the monograph of Engler and Irmscher published in 1916, not because this is necessarily now the last word but because it provides a reasonably firm and workable base which is widely used. They divide the genus into the following fifteen sections.

**Section 1**    **Micranthes** (Haw) (formerly Borophila). Most of the species in this section come from the cold northern latitudes. Our native *S. stellaris* is typical.

A number of the species in this section have been brought into cultivation and have not proved difficult to grow. They are, however, mostly short-lived plants requiring frequent renewal from seed. The leaves are usually long and nearly entire. The flowers are small and not very showy. This is not a section of any great horticultural significance.

**Section 2**    **Hirculus** (Haw) Tausch. The section takes its name from *S. hirculus* which is its most typical and widespread species.

The species in this section all like bog or damp scree conditions and will not succeed without plenty of moisture at the roots. One or two species are of horticultural interest in the bog garden.

**Section 3**    **Robertsoniana** (Haw) Ser. The section is named after Benjamin Robertson (d. 1800) a wealthy amateur of Stockwell who left his whole fortune by will (sadly for the science of botany, annulled by Lord Chancellor Eldon) to trustees to form a botanic garden of vast extent. The familiar 'London Pride' belongs to this section and is typical of the six or thereabouts species making up this well-marked group. They are valuable garden plants, and are not to be despised for their ease of culture. They prefer moist shady positions but can also be grown elsewhere.

**Section 4**    **Miscopetalum** (Haw) Engl. The name derives from the Greek *miscos* (stalk) and *petal* referring to the stalked petals as seen typically in the flowers of *S. rotundifolia*, one of the few species introduced into cultivation. They are pretty little plants, but the flowers are small and not very showy so that the section has not got a great horticultural significance.

3

**Section 5**    **Cymbalaria** Griseb. The section takes its name from *S. cymbalaria*. L. which is typical. The epithet derives from the Latin *cymbalus* meaning a cymbal, a reference to the shape of the leaves. They are annuals and of little horticultural significance.

**Section 6**    **Tridactylites** (Haw) Engl. The section takes its name from *S. tridactylites*. L., our native Wall Saxifrage. The epithet derives from the Greek words *tria* (three) and *dactylos* (finger) alluding to the three-lobed leaves. They are annuals and of little horticultural significance.

**Section 7**    **Nephrophyllum** Gaud. The name derives from the Greek *nephros* (kidney) and *phyllon* (leaf) an allusion to the shape of the leaves. Many, though not all, of the species bear bulbils at the base of the stem or in the leaf axils. Some species shed their leaves completely in the winter. *S. granulata*, the Meadow Saxifrage, is a very wide-spread and typical member of this section, which contains a number of species of interest to the horticulturist, particularly to the specialist grower.

**Section 8**    **Dactyloides** Tausch. The name derives from the Greek *dactylos* (finger) and *eidos* (appearance), an allusion to the deeply lobed leaves. The well-known Mossy Saxifrages belong to this section and it is one of considerable horticultural interest. These ubiquitous and highly decorative garden plants are mostly hybrids and cultivars which have ousted the original species. Our native *S. hypnoides* and *S. rosacea* are typical of the section. There are also a number of species such as *S. exarata* which are of considerable interest to specialist growers.

**Section 9**    **Trachyphyllum** Gaud. The name derives from the Greek *trachys* (rough or jagged) and *phyllon* (leaf), an allusion to the bristly-margined leaves. The species are mat-forming, fairly easily grown and not highly decorative. *S. aspera*, which is widespread throughout the Alps, is typical. The section has no great horticultural interest.

**Section 10**    **Xanthizoon** Griseb. The name derives from the Greek *xanthos* (yellow) and *aizoon* (an old generic name for an evergreen plant probably *Sempervivum*) but here referring to *S. aizoides*. L. because of the yellow flowers. *S. aizoides*, wide-spread throughout the Northern

4

Hemisphere and a native of this country, is the solitary species in this section.

**Section 11**   **Euaizoonia** (Schott) Engl. The name derives from the Greek *eu* (right or genuine) and *aizoon* (evergreen as above). These are the rosette-forming, silvery-encrusted saxifrages. Most of the species are easily grown and are highly decorative both in respect of their foliage and their handsome inflorescences borne in prominent panicles in the summer. *S. aizoon* or, as it is now more correctly called, *S. paniculata*, is typical of this section which is of considerable horticultural interest. Members of this section are often referred to as 'Silvery' or 'Encrusted' Saxifrages.

**Section 12**   **Kabschia** Engl. The section is named after the unfortunate Wilhelm Kabsch (1835-1864), a German botanist who fell over a cliff while carrying out phytogeographical studies in the Alps. The species are densely tufted and form cushions or hummocks. The section includes the most beautiful of all saxifrage species. *S. burseriana* (sometimes written *burserana*) is typical of this section.

The so-called 'Engleria' Saxifrages are sometimes treated as a separate section and sometimes as a subsection of the Kabschias. The latter treatment is adopted here for reasons which will be explained in due course. *S. grisebachii* is a typical 'Engleria.'

Almost all the species in this section are of considerable horticultural interest and value.

**Section 13**   **Porphyrion** Tausch. The name derives from the Greek *porphyrios* (purple) referring to the colour of the flowers. The species are all creeping, mat-forming plants and are typified by our native *S. oppositifolia*. The section includes four species only and *S. oppositifolia* is the only one of any horticultural importance but it is particularly valuable as a species of many varying and localised forms.

**Section 14**   **Tetrameridium** Engl. The name derives from the Greek *tetra* (four), *meros* (part) and *idion* (characteristic or peculiarity), an allusion to the four-partite formation of the flower. All the species are natives of Asia and only one has so far been introduced into cultivation, *S. nana*. The section has so far no horticultural significance, and with the more recent discovery of intermediate species, this Section is now usually merged with Kabschia.

5

**Section 15** **Diptera** Borkh. The section was so named by Borkhausen because the flowers with their two typically enlarged petals somewhat resemble flies of the group Diptera. The species all come from Japan or China and some are of suspect hardiness. When planted in a sheltered position some of them, particularly *S. fortunei*, can be very useful and decorative autumn-flowering plants. Only six species are in cultivation. *S. stolonifera* (syn *S. sarmentosa*) 'Mother of Thousands' is very commonly seen as a room plant.

To recapitulate it would, I feel, be a fair summing-up to say that the Genus contains three sections (DACTYLOIDES, EUAIZOONIA and KABSCHIA) of outstanding horticultural importance, six sections (HIRCULUS, ROBERTSONIANA, CYMBALARIA, NEPHROPHYLLUM, PORPHYRION, and DIPTERA) of definite though limited horticultural importance and six sections (MICRANTHES, MISCOPETALUM, TRIDACTYLITES, TRACHYPHYLLUM, XANTHIZOON and TETRAMERIDIUM) of very small or negligible horticultural importance.

---

## KABSCHIA SECTION

In commencing with the Kabschia saxifrages I am perhaps choosing a Section which has had more justice done to it than most in the *Bulletin*. It is also a section of immense ramifications when all its hybrids are taken into account. It is true, however, to say that the hybrids have usually monopolised the stage to the virtual exclusion of the species. In this present treatment I propose to deal in some detail with the true species and their natural varieties and hybrids, and only thereafter to deal (in a rather selective manner) with the best of the man-made hybrids. I shall treat the section according to its wider definition, that is including the so-called 'Engleria' saxifrages as well as the typical Kabschia saxifrages. Whilst the two types are so distinct in appearance, breeding has shown that they cross with each other with the greatest facility, providing a continuous range of intermediate forms. Even in the wild they cross readily (cf. *S. aretioides* and *S. media* in the Pyrenees) and the resultant hybrids are usually at least as vigorous as their parents.

6

*Saxifraga media* (p. 14)                    *Photo: W. Harding*

*Saxifraga grisebachii* (p. 15)               *Photo: R. Elliott*

*Saxifraga burseriana* 'Gloria' (p. 24)                    *Photo: W. Harding*

8

*Saxifraga diapensioides* (p. 25)                    *Photo: Miss D. Holford*

*Saxifraga marginata* var. *rocheliana* (p. 18)   *Photo: H. S. Wacher*

9

*Saxifraga scardica* (p. 21)   *Photo: Downward*

*Saxifraga spruneri* (p. 22)                    *Photo: R. Elliott*

10

*Saxifraga caesia* (p. 22)                    *Photo: W. Harding*

Over eighty species of this section occur on the chain of mountain ranges in the northern hemisphere which stretch from the Pyrenees through the Alps, the Balkans and Carpathians, the Caucasus, the Hindu Kush and the Himalayas to the mountains of some of the provinces of South China. However, of the fifty or so species recorded from the Far East only two appear now to be in cultivation. If any reader knows of any other Far Eastern species in cultivation I should be most grateful to hear of it. Thus almost all the species in cultivation in our gardens are of European origin even though one of the two Far Eastern representatives has been used to impart its rich colour to some of the best hybrids.

The Kabschia saxifrages vary in their ease of cultivation. It is, I feel, necessary to make a distinction between the small number of species and hybrids which are reliable plants for the rock garden and the far larger number of species and hybrids which are best wintered in a frame or alpine house. This protection is not required because of any lack of hardiness but because they prefer, when not actually growing, to be kept only just moist; also, the flowers will remain beautiful far longer if protected from rain and soil splashing. They flower so early in the year, mostly during February and March, that it is only in the Alpine House that one can enjoy their beauty to the full.

Growing the plants on the benches in the greenhouse, preferably bedded in plunging material to reduce the frequency with which watering is needed in summer and to save the roots from extreme fluctuations of temperature in winter, not only allows one to appreciate the beauty of their flowers but also makes one fully conscious of the fascination of their form and foliage. Although individual varieties may only remain in flower for three to four weeks, their silvery-green cushions have both beauty and interest throughout the year. Some kinds have an almost prostrate habit of growth whilst others make hemispherical domes or irregularly billowing mounds of tightly-packed rosettes which look particularly well engulfing pieces of tufa rock. The leaves are usually firm, in some cases even spiny, and the individual rosettes are objects of extreme beauty with their silvery encrustations and often prominent patterns of lime glands.

The number of species which can be considered reliable rock garden plants varies somewhat with the nature of the soil and its drainage. In general however only the following should be relied upon for this purpose:- S. 'Elizabethae', S. 'Haagii', S. 'Jenkinsae', S. x *apiculata* and *apiculata* var. *alba* and S. *sancta*. These are not difficult if attention is given to their placing and planting. These easier sorts should, I consider, be planted in the open with full exposure to the sun and, provided that they do not get dried out

11

at the root, this treatment will ensure the greatest floriferousness. Most of these plants are capable of making broad mats of foliage bearing quantities of bloom and are distinctly ornamental in the rock garden.

The larger number of species and hybrids will not stand such treatment and should be given pan culture and the protection of a frame or greenhouse from the end of September to the end of April. They are, with only one or two exceptions, very far from being difficult plants provided their essential needs are met. These are (1) a well-drained gritty compost (2) an intelligent watering regime throughout the year and (3) shading from fierce sunlight.

The essential attributes of a good compost for Kabschia saxifrages are, as mentioned above, a gritty open texture and the ability to drain rapidly. The mixture I use, though no doubt many others serve equally well, consists of 1 part loam, 1 part fine peat, $\frac{1}{2}$ a part silver sand, and $\frac{1}{2}$ a part coarse grit. To the whole I add a dash of John Innes Base Fertilizer and the pans are top-dressed deeply with limestone chippings. The compost, if of a suitable texture, should, after being clenched tightly in the hand, fall readily apart when the pressure is released. Water applied at the top of the pot should rapidly find its way through to the drainage hole. At the bottom of the pot perforated zinc plus a generous layer of crocks or coarse Cornish Sand will ensure drainage of excess water. In the case of *S. lilacina* the compost must be lime-free and topped with granite and not limestone chippings.

The next cultural requirement is correct watering at all times. Once growth has started in the Spring, plenty of water is required and if the compost and drainage are correct, it will hardly be possible to overwater them. The growing period comes quite early in the year and in many cases flowering and the formation of new rosettes of foliage take place simultaneously. Immediately after flowering is a time of particularly rapid growth. Later in the season when growth slows down, water should be given less freely and after the plants have been brought under cover in late September, watering should be progressively reduced to a minimum so that during November and December they are kept barely moist. They should not however on any account be allowed to dry out completely.

The third requirement concerns the correct provision of shading throughout the year. During the winter they should be allowed the fullest amount of light that can be given them, but during the summer full exposure to fierce sunlight can easily produce severe scorching of the whole plant which may be fatal. If plunged out during the summer, they are best placed where they will get only the early morning sun, and even then slatted shading may be

12

advisable. If kept inside during the summer, fairly heavy shading is advisable.

Kabschia saxifrages are easy plants to propagate and it is both interesting and wise to maintain a flow of young plants which can always be found a home if not needed for growing on. Most writers advise the taking of cuttings in May and June after flowering, but I prefer to take them as soon as the plants have started to produce new growth which with me occurs during the month of March. This gives the cuttings, which root in about six weeks, a useful growing season before they are potted up in the early autumn or subsequent spring. The compost I use for cuttings consists of 1 part loam, 1 part fine peat, 1 part silver sand and 1 part grit. I usually line out the cuttings in 9 inch square pans which each take 72 cuttings comfortably and cover them with sheets of glass. Any sized pan is however suitable according to the numbers taken. The cuttings, each of which is about ½ an inch long and consists of a single rosette, are a little fiddly to prepare and handle, but that is all. The lower leaves should be carefully removed so that the soil can be pressed firmly against the stems. September/October is also a suitable time.

Having dealt in general with the cultivation of Kabschia saxifrages it is now necessary to return to the recognition and treatment of the individual species. After some hesitation I have decided, purely for purposes of grouping, to break down the Kabschia Section into the seven sub-sections of Engler and Irmscher but not to detail the botanical characters, mostly of a rather minute nature, on which the sub-sections have been defined. These sub-sections and the species which fall into them are as follows:-

(1) MEDIAE — The 'Engleria' saxifrages. *SS. media, porophylla, grisebachii, stribrnyi, sempervivum (thessalica), luteo viridis.*
(2) JUNIPERIFOLIAE — *SS. juniperifolia, caucasica, sancta.*
(3) KOTSCHYANAE — *S. kotschyi.*
(4) MARGINATAE — *SS. lilacina, marginata, scardica, spruneri.*
(5) SQUARROSAE — *SS. caesia, squarrosa.*
(6) RIGIDAE — *SS. burseriana, diapensioides, tombeanensis, vandellii.*
(7) ARETIOIDEAE — *SS. aretioides, ferdinandi-coburgi.*

## Part I    Kabschia Species

### Mediae

There are six species belonging to this sub-section in cultivation

13

in this country. Five have red flowers whilst the other has yellow ones. The former, which are the more usually grown species, all produce elegant crozier-shaped inflorescences in varying shades of red from the centres of rosettes of symmetrically arranged silvery leaves.

The colour is given by the velvety pile of closely-set red glandular hairs that cover the whole inflorescence and stem. The baggy red calyces are the most prominent part of the flower. It is while the inflorescence is still in its crozier stage and before it unfurls itself that it is at its most attractive and the colour most intense. The plants are very attractive for two full months and even at other times the silvery rosettes of leaves always evoke appreciative comments. All may readily be grown from seed.

The cultivated species are all inhabitants of Europe, but in general they tend to have their own separate ranges.

### S. MEDIA (illustration page 7)

This occurs only in the Pyrenees and is readily distinguished from the other species both by the character of the leaves and the inflorescence. The rosette leaves do not lie nearly as flat as in the other species but stand out at an angle of approximately 45 degrees and then recurve outwards towards the upper half of the rather sharply pointed leaf-tips. The inflorescence forms a more compact spike than I have seen occurring in the other species and as occurring in the Pyrenees; it is unbranched, contrary to some written descriptions. I have not found S. *media* by any means the easiest to keep in cultivation and, but for the precaution of rooting some side rosettes as soon as I obtained it, I should now be without it. It is however a satisfying and distinctive plant when seen at its best, and is therefore worth persevering with.

### S. POROPHYLLA

Proceeding eastwards to the Apennines in Italy, the territory of S. *porophylla* is encountered. The species is particularly common and widespread throughout the Abruzzi Mountains where it grows, both in rock fissures and in scree, in company with SS. *callosa* (*lingulata*), *paniculata* (*aizoon*) and *moschata*.

Both in the length of the leaf

and of the inflorescence it is markedly shorter than *S. media* and it makes a miniature and most attractive plant. The inflorescence is unbranched and about 3 inches high. The rosette leaves lie very flat and are somewhat spatulate in shape.

This also is not the easiest of plants to cultivate although I still have it thriving from a collection made in the Abruzzis five years ago. It has even produced some self-sown seedlings and has over-wintered outside for three years planted in a trough.

S. SEMPERVIVUM. (*S. thessalica* or *S. porophylla* var. *thessalica* of catalogues.)

This species is spread fairly generally over the mountains of the Balkans. A number of sub-species are recognised, but I will not attempt to describe these as their taxonomy is difficult. In general however the leaves are narrower and more needle-like than the type species, with unbranched flower spikes about 4 inches high. A somewhat broader-leaved sub-species, almost like a miniature *S. grisebachii*, occurs in Montenegro. *S. sempervivum* is not a difficult

species to grow with the protection of the Alpine House. A cultivar 'Waterperry' is of similar habit but with the inflorescence mauve in colour rather than reddish-purple.

S. GRISEBACHII (illustration p. 7)

On the other side of the Adriatic in Albania and the adjoining areas of Greece, Bulgaria and Yugoslavia, *S. grisebachii*, the most showy and horticulturally useful of these species, is found. The type is only rarely seen in cultivation, having been almost completely supplanted by its cultivar 'Wisley' which is more brightly coloured and vigorous. The rosettes are 1½ to 2 inches across and the individual leaves are 1/5 inch wide. The latter are very silvery and laid flat in a beautifully intricate pattern. Very early in the year a red boss appears in the centre of each flowering rosette, which progressively lengthens into a velvety red inflorescence about 6 inches long.

As these 'Engleria' species go, it is not a difficult plant to grow with the protection of the Alpine House.

## S. LUTEO VIRIDIS

This, the only species in cultivation with yellow and green flowers, hails from the Carpathians and Bulgaria, though its range also extends into Asia Minor. The yellow petals are enclosed in a green corolla and the flowers are borne in a terminal panicle rather like *S. stribrnyi*. It is a pretty little plant and can be grown readily, if slowly, from seed, but it never makes a large or spectacular plant. It is unlikely to persist if grown in the open in winter. In summer it needs more shade than its relatives.

## S. STRIBRNYI

This, the last of the red-flowered species, occurs in northern Greece and Bulgaria and is readily recognised by the very branching paniculate inflorescence. It is a handsome plant, well established in cultivation and not temperamental. I have, however, never tried growing it out-of-doors.

## Juniperifoliae

This is not, horticulturally, a very valuable section but nevertheless it does contain two species *S. juniperifolia* and *S. sancta* which are fairly widely grown, though the popularity of both species is not very great and appears to be further declining. A third species *S. caucasica* var *desoulavyi* is also still in cultivation though seen infrequently. These species are all from the Caucasus, though in the case of *S. sancta* the range extends through Asia Minor and the Eastern Balkans.

Another four or five species, all from the Caucasus, have either not been introduced into cultivation or have only remained fleetingly in cultivation. S. artwenensis usually has white flowers.

## S. CAUCASICA VAR. DESOULAVYI

This is quite an ornamental plant with small yellow flowers borne singly on 2 inch flowering stems. The foliage forms mats of dark green rosettes. The leaves are acutely pointed. It was introduced by Sündermann but has never achieved any great popularity. It resembles *S. juniperifolia* closely, but flowers more freely in cultivation.

## S. JUNIPERIFOLIA

This is a disappointing species, though readily obtainable. The leaves, when crushed, are said to smell of Juniper, though I personally have never been able to detect this. They are bright green in colour and form a cheerful and ornamental mat of rosettes but unfortunately the floral display is usually sparse in this country, and even when allowed to become somewhat pot-bound I have

16

rarely had enough flower to make it a really effective plant. When it does flower the inflorescences, which are made up of 7 or 8 flowers on a 2 inch flowering stem, are not very spectacular, as the narrow petals make the stamens of the flower its most noticeable feature.

## S. JUNIPERIFOLIA VAR. MACEDONICA

This is a somewhat larger growing and more decorative plant, grown in the past as *S. macedonica*. The flowers are broader-petalled and more decorative than the type and are set on 3 inch stems. *S. juniperifolia* is a variable plant and this is one of its best forms.

## S. SANCTA

This is a long-standing inhabitant of our gardens and has proved itself persistent even under border conditions. Of recent years it has become more difficult to find as people have become tired of its shy flowering habits and discarded it for more floriferous and equally hardy hybrids such as *S. x. apiculata* and *S.* 'Elizabethae'. However *S. sancta* is not shy flowering if grown under starvation conditions and it is the richer fare that it is often given which inhibits its flowering. Planted on a hard rubbishy soil it can be covered in flowers and this should be viewed as an asset to be exploited. Three to 7 golden-yellow flowers are borne in a globose inflorescence on a 2 inch stem. The dark-green leaves are about half an inch long, toothed and ending in a spine. The plant is mat-forming. It is interesting to note that in *Flora Europaea*, Professor Webb shows it as a sub-species of *S. juniperifolia*, but this is a very recent development.

## Kotschyanae

This is the sub-section of least horticultural interest. It includes only two species, one of which, *S. meeboldii*, is a Himalayan species not in cultivation. The other species, *S. kotschyi*, is in cultivation and is described below.

## S. KOTSCHYI

This species comes from Asia Minor, from the regions of the Pontus and Taurus mountains. Although it has been in cultivation for many years it has no particular merit. The few yellow flowers are not large, or of very impressive appearance, and are borne in 1½ inch lax inflorescences. Even where this plant is grown, it is somewhat doubtful whether it is in fact the true species which may be even less horticulturally interesting than the plant often seen under this name.

The foliage makes a flat mat with blunt encrusted leaves.

17

## Marginatae

This section includes four species which are in cultivation, all originating from southern and central Europe except *S. lilacina* from the Himalayas. There are also at least fourteen other species in the Far East which are not yet in cultivation.

### S. LILACINA

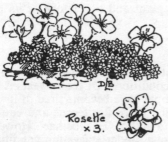

Rosette × 3.

This Himalayan species is of outstanding importance as the source of the colour in such famous hybrids as 'Jenkinsae', 'Myra', 'Cranbourne' etc. Its flowers are a true lilac and sit almost stemless on a flat mat of hard grey-green foliage. It is unique, not only in its colour, but also because it is a lime-hater. It prefers to be shaded from all direct sunlight and it requires plenty of water during the growing season, allied with sharp drainage. Whilst not a difficult plant it is not easy to keep in an ornamental condition unless these directions are followed closely. It flowers at the end of March and at the beginning of April in the south.

### S. MARGINATA (illustration—of the variety *rocheliana*—page 9)

This variable species ranges through Southern Europe from the Apennines to the Balkan mountains and the Carpathians. The handsomest and most free-flowering form, which is the type plant, is found mainly in Southern Italy. It even occurs on the highest point of the Sorrento peninsula at an altitude of only 1400 metres. Three named varietal forms are in cultivation. These are var. *rocheliana*, var. *coriophylla* and var. *karadzicensis*. The first has the longest leaves in a rather flatter and less columnar rosette and the second has smaller leaves arranged in a distinctly columnar rosette; they both occur over the same wide range of the Balkans and the Carpathians. The third, var. *karadzicensis*, is quite distinct in that it is a very dwarf and compact form with tiny columnar rosettes and heads of few flowers on stems only an inch in length. It makes a fascinating slow-growing pan plant and received the Award of Merit in 1940. It was re-discovered and re-introduced from Albania before the war by members of the Alpine Garden Society.

*S. marginata* in its more floriferous forms is a first-rate garden plant with handsome inflorescences of four, five or more large,

18

*Saxifraga* x *biasolettii* 'Crystalie' (p. 28)

*Photos: R. Elliott*

*Saxifraga* 'Valerie Finnis' (p. 33)

*Saxifraga* 'Winifred' (p. 33)                    *Photo: R. Elliott*

20

*Saxifraga biternata* (p. 41)                    *Photo: W. Harding*

slightly funnel-shaped pure white flowers borne on two-inch flowering stems. The sepals are reddish-purple and before the flowers open the contrast is most attractive and striking. The petals are of good substance. It is not difficult to grow, though I believe that it prefers to have its roots somewhat cramped in close proximity to rock to produce the best flowering display. Too much exposure to the sun should be avoided. In the south it flowers at the end of March and the beginning of April.

## S. WENDELBOI

This species from Iran is close to S. marginata. It was shown by Archibald in March 1973. Rosettes flat and silvery, leaves rather broad and bluntly pointed, flowers white, 2 or 3 on 1-inch stems, calyx red.

## S. SCARDICA (illustration page 9)

This species hails from Mt. Olympus and the adjoining mountain ranges of Albania, Greece and Jugoslavia. It is less frequently encountered in cultivation than its so-called variety *obtusa*, which is an easier but less handsome plant.

The true *S. scardica* is a plant of very characteristic appearance with relatively large rosettes of broad spiky leaves making a hard compact dome. It bears an inflorescence of from three to eleven pure white flowers on 3-inch to 4½-inch reddish stems and is highly decorative when seen in full flower. Strains with pink-flushed flowers also occur in the wild alongside those with pure white flowers. I do not find it an easy plant to grow as it seems very susceptible to scorch. It is, I have found, necessary to give it even more shade than most other Kabschias except possibly for *S. lilacina*.

Its variety *obtusa* differs so markedly in appearance, with its narrow light green non-spiny leaves and its 3-inch long green flower stems bearing 3 to 6 white flowers, that I find it difficult to believe that it is a true variety of the species and not a hybrid. Engler and Irmscher do not record any localities for it and in the *Botanical Magazine* it is stated that the Kew plants from which it was described were received from the firm of Sündermann. It is a much easier plant to grow than the species and will take more sun without scorching. It is normally very floriferous but the petals do not seem to have the same substance as some of the other Kabschia species and tend to twist and flop giving the flower a slightly bedraggled appearance.

There is said to be a red-flowered variety *erythrantha* growing on Mt. Kyllene in the Peloponnese, but if it exists it has not been introduced. The var. *erythrantha* of commerce is not a true red-flowered variety but only a pink flushed form of the normal species.

S. SPRUNERI (illustration page 10)

This species, hailing from Albania, Northern Greece and Bulgaria, is readily recognisable as being the only Kabschia with distinctly hairy foliage. It is not very commonly grown in spite of being a decorative plant with clusters of pure white flowers set on short stems. I have not grown it myself and can only presume that its hairy nature makes it a rather more difficult plant to keep through the winter than most of the Kabschia species. Its seed is often offered in the Alpine Garden Society's Seed Distribution so that it is firmly in cultivation.

## Squarrosae

This small section includes but two species, *S. caesia* and *S. squarrosa*, both of which are found only in Europe. There is also a naturally occurring hybrid between them called S. x *tyrolensis*.

S. CAESIA (illustration page 10)

Of the two species this is far the more widespread, occurring as it does on all the main mountain ranges of Southern and Central Europe from Spain to Poland. Where it occurs, it is usually an extremely common plant. It is particularly partial to fissures on isolated rocks where its roots delve deep down into the rock to a steady supply of moisture. Seeing it growing under these conditions one would not think that it was a plant that presented any greater difficulties of cultivation than *S. aizoon*. Unfortunately this is not the case and *S. caesia* is rarely seen at the Alpine Garden Society's Shows from which one must assume that it is not particularly easy to retain in cultivation in this country. I have flowered it, propagated it and lost it. At the present time I have a number of promising plants growing against and on Tufa rocks and I feel that this is the surest way to grow it. I do not believe that, under our different conditions in this country, it is safe to give it the full sun in which it revels in its native haunts. Yet it needs a reasonable amount of illumination if it is to be kept compact and in character. During the winter it should be watered very sparingly.

22

Rosette ×3

S. caesia

The rosettes are very small, being only about an eighth of an inch in diameter and flat. This results from the leaves being recurved outwards from their bases whereas *S. squarrosa*, the next species, is distinguished only by the leaves recurving in their end portions so that instead of a flat rosette a 'squarrose' or somewhat columnar and spiky rosette is built up. *S. caesia* is also distinguished by the fact that only the upper half of the floral stem is usually glandular-hairy whereas in *S. squarrosa* it is usually the lower half which is glandular-hairy. These however are details to settle the identification in case of doubt. In point of fact the two species once seen give quite distinct and separate impressions. The rosettes of *S. caesia* are a blue-green covered with lime encrustations which produce a grey-green overall effect.

The flowers, borne on two to three inch flowering stems, are surprisingly large and handsome in relation to the minuteness of the rosettes and are pure white in colour, two to four flowers being borne to a stem. They appear very late in the season for a Kabschia saxifrage, not being produced until June. This is a very neat and decorative plant when well grown and deserves to receive more attention.

Rosette ×4

S. squarrosa

## S. SQUARROSA

Reference has already been made to the features which distinguish this species from the foregoing. It has a far more limited distribution, occuring as it does only in Italy, Austria, and Jugoslavia—and there only in the area of the Eastern Alps which straddles the junction of the three countries. It is particularly common in the Dolomites. It is a rare plant in cultivation where it is liable to be

23

lost during the winter. This species is also probably best grown on Tufa. The flowers and flower stems are very similar to *S. caesia* and also are not produced until June.

## S. X TYROLENSIS

This natural hybrid of *S. caesia* and *S. squarrosa* occurs not uncommonly where the two species grow side by side. It is somewhat more amenable to successful cultivation than either and is not infrequently seen on the show bench where it displays more vigour than either of its parents.

## Rigidae

This sub-section includes four species in cultivation, all of which come from within a restricted area of Europe in Northern Italy and the immediately adjoining areas of the neighbouring countries. *S. burseriana*, the commonest of the four species, is very significant horticulturally both in its own right and as a parent of many fine hybrids.

## S. BURSERIANA (illustrated on p. 8 in its variety 'Gloria')

This species is spelt *S. burserana* in the recently published Flora Europaea and so this must be presumed to be its valid name. However as *S. burseriana* is the spelling used to date in all previous horticultural and botanical publications this will be retained for the purposes of this article.

*S. burseriana* is, to my mind, the gem of the Kabschia Section. It occurs in the Eastern Alps, the Dolomites and the Karawanken and varies considerably in form over its geographical range. Typically the leaves are glaucous grey-green, stiff, spiny, and rather erect. The pure white flowers are borne on 2 to 3-inch reddish stems. The broad petals spread out horizontally making a delightful flower of large size and classic form with slightly overlapping petals. In general it appears that the plant is smaller in all its parts at the Eastern end of its range than at the Western end but this distinction may not hold universally true.

The Western form is found growing in the mountains in the vicinity of the town of Trento and hence is called var. *tridentina*. This appears to be the same plant that has been called var. *major*. A particularly choice selection from this race was made by Farrer who distributed it under the cultivar name of 'Gloria'. It has flowers of over an inch in diameter and also has the very great merit of not being at all a shy flowerer.

Var. *minor* is a higher altitude plant with much tighter smaller rosettes and smaller flowers on 1 inch pedicels.

The type species, as I have collected it near Tarvisio in Italy, approximates more closely to var. *minor* than to var. *tridentina* but is not so compact or small in the rosette and carries quite large flowers of perfect shape on 2 inch long red pedicels. I found it at an altitude of only 2500 feet which shows that on occasion it will descend to quite modest heights. The species has I believe the reputation of being shy-flowering in cultivation, but I have had it flowering profusely when in good health though more sparsely when not so obviously in good health. I must admit I have not found it an easy plant to keep in good health as chlorosis has tended to occur during the summer. I think it favours rather more shade than some Kabschias, allied with very sharp drainage, and these are the two points which the cultivator is well advised to watch. It may be questioned why one should go to this trouble when 'Gloria' or its near allies 'Brookside' and 'Crenata' are so much easier to grow. I can only say that I find the typical species a particularly attractive plant when seen at its best. It also proves to be a more successful seed-parent than the others. I have never yet obtained seed from 'Gloria' and it would be interesting to know whether they have the same chromosome numbers. For the average grower I am sure that 'Gloria' and 'Brookside' are the best cultivars to grow.

There are a number of other plants, some yellow-flowered, named as varieties of *S. burseriana* which are almost certainly hybrids. These will be dealt with later under Kabschia hybrids.

## S. DIAPENSIOIDES (illustration p. 8)

This species occurs in the South Western Alps, in Italy and Switzerland. It is a very tight little plant with closely huddled small rosettes of short leaves and is very slow-growing. It bears 2 to 6 pure white flowers on a 2 inch flowering stem. In cultivation it tends to be rather shy flowering unless particularly well suited. It is a difficult plant to grow and, when not actively growing, it can very easily be lost from over watering. Winter damp appears to be the danger. I have seen good flowering plants grown in double pots with watering during the winter confined to the material in the outer pot. I must admit to having lost my only plant last winter after three years.

*S. diapensioides* 'Lutea' is an easier plant to grow but has the appearance of a hybrid rather than a variety.

## S. TOMBEANENSIS

This species is found both on Monte Baldo to the east of Lake Garda and on Monte Tombea to the west of the same lake. It is very localized in its occurrence. Fortunately it is readily available in commerce.

It forms a hard compact hummock of small green rosettes in which the short leaves are tightly pressed inwards. Whilst not difficult to grow, it is very shy of producing its flowers and it has only so far flowered once with me. The two-inch flowering stems bear two to three large pure white flowers. I have now planted young plants in holes in a tufa rock, and shall be interested to see whether it is any more floriferous grown in this manner. It is easily propagated from cuttings.

## S. VANDELLII

This species occurs in limestone cliffs in the Italian Alps but I have not seen it in the wild. I have, however, cultivated plants which flower well; its leaves are considerably longer than the two foregoing species, being about ½ inch long, and it grows into an exceptionally hard cushion. 3 to 8 pure white flowers are borne on a 2 to 3 inch flowering stem.

## Aretioideae

This small sub-section is composed of but two species, *S. aretioides* and *S. ferdinandi-coburgi*, both yellow-flowered, a factor which has proved of some importance to horticulturists.

## S. ARETIOIDES

This species is confined to the Pyrenees. It forms hard compact uneven hummocks, bearing few-flowered inflorescences of golden-yellow flowers set on very short stems. It hybridizes readily in nature with *S. media* (an 'Engleria' saxifrage of the Mediae subsection) which grows in the same region.

*S. aretioides* is not an easy plant to grow, nor is it a very decorative species as the flowers are small and not very showy. It has, however, proved a most useful species for hybridization and has imparted its yellow colour to many good hybrids where the other more decorative characteristics derive from such species as *S. burseriana* and *S. marginata*. It does not like much direct sun.

## S. FERDINANDI-COBURGI

This species is found in its typical form in the Pirin Mountains in Bulgaria. It is a plant with tiny congested rosettes and very short leaves. The golden-yellow flowers are borne in cymes of 7 to 12 flowers on stems 1 to 2 inches long. The sepals are green.

The species has however two superior varieties, var. *radoslavoffii* which, I am told by Herr Schacht, is found on Ali Botusch in Macedonia, which is south of the Pirin Mountains, and var. *pravislawii* from the Rhodope Mountains to the east. The former is more vigorous than the type and has larger flowers with very narrow calyces. The latter, which is the finest form of all, is a very vigorous plant, free-flowering with deep golden-yellow flowers and

26

broad reddish-brown sepals. The inflorescences contain up to 20 flowers borne on a 3 inch flowering stem and these make it as decorative as any of the Kabschia hybrids when well grown. The rosettes of leaves are four times as large as in the type.

The species and its varieties all have handsome glaucous grey-green foliage which is always extremely decorative even when not in flower.

## The Far Eastern Kabschia Species

As well as the solitary Himalayan species, *S. lilacina*, which has already been referred to under the Sub-section Marginatae, more than fifty other Kabschia species inhabit the Sino-Himalayan mountain ranges, though so far hardly any of them have been brought, even fleetingly, into cultivation. It is difficult to refer many of these with exactitude to particular Sub-sections, as Engler & Irmscher's subdivisions were mainly constructed around the European species, only a few of the Far Eastern species having been discovered at that time. Though fifty four species are now recognized from this area many of these are of very local occurrence and only marginally distinct.

In addition to *S. lilacina*, there is one other Far Eastern species, *S. andersonii*, which now seems to have come to stay with us. This was given a Preliminary Commendation when shown by Mrs. Crewdson of Kendal at the Spring Show in 1959 under the erroneous name of *S. stolitzkae*, which was subsequently corrected. In 1964 it was shown again by the same lady and received an Award of Merit under its correct name. It forms flat cushions of small lime-encrusted rosettes up to 1.5 cms. in diameter carrying 5-7 cms. inflorescences bearing up to 5 white or pale pink flowers. It was introduced from Nepal by Col. D. G. Lowndes in 1950. It is now available commercially. I have found it a rather shy flowerer.

Another species of which a photographic record has appeared in the *Bulletin* (Vol. 38 page 175) is *S. lowndesii* which has almost sessile solitary mauve-pink flowers, said to be superficially reminiscent of *S. oppositifolia*. It was introduced fleetingly into this country by Col. D. G. Lowndes in 1950 but did not persist.

During the period between 1970 and 1980, thanks mainly to the devoted efforts of George Smith and Duncan Lowe, further Far Eastern species have been brought into cultivation and seem likely to remain with our more skilful growers—even perhaps being used for hybridization. *SS. geogei, hypostoma, pulvinaria* (syn. *imbricata*), *quadrifaria* and *stolitzkae* are at present the main species involved and of these *SS. hypostoma* and *stolitzkae* are the most attractive as well as proving fairly tractable given cultivation in a sharply drained medium with some overhead winter protection. *S. pulvinaria* is described and illustrated in *Bulletin* Vol. 42, pp. 204

27

and 207, whilst *SS. hypostoma* and *stolitzkae* are described and figured in considerable detail in *Bulletin* Vol. 45, p. 276 and Vol. 46, p. 140. *SS. afghanica, chianophila, kumuanensis, likiangensis, meeboldtii, pulchra* and *rupicola* are still awaited.

## Part II   Kabschia Hybrids

So far we have only dealt with the species and a handful of the commonest natural hybrids. We will now look at the multiplicity of man-made hybrids. As these are so numerous I feel that it will be convenient to deal with them in three broad groups even though the placing of a few of them may be tricky to decide. These groups are:
  1. 'Engleria' crossed with 'Engleria'.
  2. 'Engleria' crossed with 'Typical Kabschia'.
  3. 'Typical Kabschia' crossed with 'Typical Kabschia'.

### 'Engleria' crossed with 'Engleria'
The following are the best hybrids of this type:—

S. X BIASOLETTII 'CRYSTALIE'. (*S. sempervivum* (syn. *S. porophylla* var. *thessalica*) x *S. grisebachii*) A.M. 1948 (illustration p. 19)

This is an outstandingly good hybrid of rapid growth and most satisfactory vigour. It makes a better flowered specimen plant than *S. grisebachii* 'Wisley Var.', though the individual inflorescence is not quite so handsome. It forms lateral rosettes with greater freedom than many of the 'Englerias' and makes a sizeable plant quickly. The rosettes are intermediate in size between the two parents and are very silvery. The inflorescences are crimson-red and about seven inches high.

S. 'GUSMUSSII'. (*S. sempervivum* (syn. *S. porophylla* var. *thessalica*) x *S. luteoviridis*)

This is a less vigorous hybrid than the previous one but nevertheless attractive with its spikes of rather paler orange-red flowers.

S. 'SCHOTTII'. (*S. stribrnyi* x *S. luteo viridis*)

This is an attractive hybrid with branching inflorescences covered with rose-red glandular hairs and small yellow flowers. The rosettes are flat and very silvery.

S. 'STUARTII'. (*S. aretioides* x *S. media* x *S. stribrnyi* ?)

This is a hybrid of complicated genealogy bearing very compact 4 inch brick-red spikes of flowers rather late in the season. As will be seen it may possibly have a little 'Typical Kabschia' blood in it.

S. 'BERTOLONII', S. 'DOERFLERI' AND S. 'FLEISCHERI'

These are other 'Engleria' hybrids less frequently seen now and with nothing very particular to offer in comparison with the above.

S. 'FEDERICI-AUGUSTI'

There seems to be some doubt whether this is a naturally occurring form of *S. sempervivum* or whether it is a natural hybrid of this species. Its rosettes are fairly large with rather broad leaves, and the rose-red inflorescences are about 4 inches high and freely borne. It makes a good pan plant.

### 'Engleria' crossed with 'Typical Kabschia'

This group of hybrids includes a number of valuable and interesting plants which may be intermediate in character or may veer towards one or the other parental type. Compared with 'Typical Kabschia' hybrids their merit lies in the extended colour range of their flowers, the length of time over which individual flowers retain their beauty, the larger number of flowers to the inflorescence and their handsome silvery foliage. Against this their flowers lack the large-petalled classic form of the 'Typical Kabschia' type, being usually less open in shape and impressing more by their collective form. The following are in my view the best of these hybrids.

### S. 'BRIDGET' (*S. marginata* var. *coriophylla* x *S. stribrnyi*)

This hybrid carries about 12 flowers of a typical pale mauve colour on each of its branching 4 to 5 inch inflorescences. The decorative rosettes, composed of silvery leaves about one eighth of an inch in width, are about ¾ inch in diameter. This is a handsome and distinctive plant of good constitution.

### S. 'CHRISTINE'

This is a dainty and attractive hybrid of unknown parentage which I have not grown. It is distinguished by its deep cerise flowers which are borne two to three to a 1½ inch green stem.

### S. 'GRACE FARWELL'

This hybrid of unknown parentage is distinguished by its late flowering habit (April-May) and by its rich wine-red flowers which are borne singly or in pairs on 1½ inch stems. The rosettes are neither as large nor as attractive as some hybrids and are rather dark green in colour. The individual flowers, though of such an unusual colour, are rather closed in form and disappointing on this account. It is an easy plant to grow and soon makes a broad mat of foliage. I mention this hybrid because of its lateness and the rich colouring of its flowers, but not everybody may feel that it is worth growing.

### S. 'HEINRICHII'. (*S. aretioides* x *S. stribrnyi*)

This is an interesting and unusual plant in that the yellow petals of the first parent combine with the reddish-purple sepals of the second parent to give bicolored flowers. 10 to 12 of these are carried in a branching yet compact 2½ inch inflorescence. The 1 inch diameter rosettes, with individual leaves over one eighth of an inch in width, are typically 'Engleria' in form, new rosettes being produced somewhat sparingly so that the plant increases only slowly in size.

S. 'KELLERERI' (*S. burseriana* x *S. stribrnyi*)

This is an excellent hybrid which ought to be in every collection. One of its most interesting characteristics is its early flowering habit. It is reputed sometimes to commence flowering before Christmas though with me it has never opened its first flowers before mid-January (I have always found S. 'Maria Luisa' the first to open its flowers in early January). It starts to flower regularly before either of its parents. The 3 to 4 inch high inflorescence usually carries three pink flowers, the clear colour of which is most attractive. The foliage, composed of 1 inch diameter silvery rosettes of a typically 'Engleria' nature, is very handsome.

S. 'KEWENSIS' (*S. burseriana* var. *macrantha* x *S. porophylla*)

This hybrid has narrow spiny grey-green leaves in compact rosettes and carries sprays of lilac-pink flowers on 3-inch reddish stems. I have not found it to be a particularly easy plant to keep in good decorative condition, for with me, the lower leaves tend to die off and mar the appearance of the plant. This may, however, simply be my bad luck.

S. 'MARIE THERESIAE' (*S. burseriana* x *S. grisebachii*)

This is a very distinctive and attractive plant which I have not grown. It bears clusters of small pink flowers on 3-inch reddish stems and rosettes of short grey-green pointed leaves.

S. 'RIVERSLEA' (*S. lilacina* x *S. porophylla*)

This hybrid forms a particularly hard and compact mound of silvery foliage. It bears 1, 2 or 3 bell-shaped flowers of a rich wine-red colour on 1½ to 2 inch inflorescences. It is a neat plant characterized by its compactness and the extremely rich colouring of the flowers and is worthy of inclusion in any collection.

S. 'SUENDERMANNII MAJOR' (*S. burseriana* x *S. stribrnyi*)

This is the last hybrid I intend to mention in this group. It is in fact my favourite and a very good hybrid indeed. It has pale blush-pink flowers of very good substance borne 3 to a 2½ inch inflorescence. The crowded silvery-green rosettes, about ⅜ inch in diameter, form a handsome mound above which the flowers stand well clear. It is an excellent grower and the flowers remain beautiful for the best part of two months.

### 'Typical Kabschia' crossed with 'Typical Kabschia'

This group contains over a hundred hybrids, including many of the best known. Since many are very similar and others have been superseded by 'better' varieties, I shall be content to describe the twenty varieties which I personally consider to be the best, realizing that such a choice is inevitably a subjective one.

S. X APICULATA. A.G.M. (*S. marginata* var. *rocheliana* x *S. sancta*)
This is a hybrid of magnificent constitution which can quite safely be grown in the rock garden with little in the way of special preparations. The foliage is bright green and the primrose-yellow flowers are borne in an inflorescence of about 10 to 12 individual blooms held well clear of the foliage on 2 inch stems. It flowers to perfection when given full exposure to the sun and only gives a mediocre performance when placed in partial shade. It has a sport *S.* x *apiculata* var. *alba* which is white-flowered and was given the Award of Merit in 1964.

S. 'ARCO-VALLEYI' (*S. lilacina* x *S. burseriana* 'Minor')
This is an old but meritorious hybrid. The large flowers, sitting on very short stems, are pale pink fading to white. They have petals of good substance and so remain decorative for longer than many of the hybrids.

S. 'BOSTON SPA'
This is a very easy, robust and decorative hybrid with dark green, spiky foliage producing its 2 to 3 inch flowering stems freely. The latter bear 3 to 4 primrose-yellow flowers of medium size.

S. BURSERIANA 'HIS MAJESTY'
This has all the appearance of being a hybrid. It has the large white flowers of *S. burseriana* but these are a fraction less open and the petals are flushed with pink at the base. The flowers, each of which shows a prominent red nectary ring, do not all mature on the same length flowering stems. It is perhaps not quite so easy to grow well as some hybrids, but is nevertheless well worth attempting.

S. BURSERIANA 'MAJOR LUTEA' A.M. 1963
This also appears to be a hybrid. It bears large sulphur-yellow flowers freely on deep red flowering stems well clear of the grey-green foliage. The form of the individual flowers is very good, with broad petals of good substance. It is a good grower and is without doubt amongst the very best of the new yellow hybrids.

S. BURSERIANA 'SULPHUREA'
This is self-evidently a hybrid. It makes a close mat of decorative grey-green foliage from which rise the 1 to 2 inch flowering stems on which are borne, singly, the relatively large, slightly nodding flowers which are sulphur-yellow in colour. It is an easy and rapid grower.

S. 'CRANBOURNE' A.M. 1936. Parentage unknown
This is a 'must' amongst Kabschia hybrids with its deep rose-pink flowers set almost stemlessly on a flat mat of dark green foliage composed of many flattened and closely huddled rosettes. The flowers unfortunately fade to a less interesting pink after about a week. It is a very reliable grower.

31

S. 'ELIZABETHAE' (*S. burseriana* x *S. sancta*) A.G.M. (1926)

This, as previously mentioned, is one of the few Kabschia hybrids which makes a thoroughly satisfactory plant in the rock garden. It spreads rapidly, grows happily in full sun and is always prodigal in the production of its flowering stems, each bearing three or four medium-sized clear yellow flowers. It is an old hybrid but should feature in every collection.

S. 'FALDONSIDE' (*S. aretioides* x *S. marginata* var. *rocheliana*)

It is with certain mixed feelings that I include this hybrid, which is an old one produced by Dr. Boyd. It is indubitably a very decorative yellow hybrid with its very hard grey hummocks of foliage and large clear yellow flowers set well above the foliage. I have not however found it to be an easy hybrid to keep for a number of years for growing into a large specimen. It seems to be more sensitive than many to exposure to full sunlight which often causes its rapid demise.

S. 'HAAGII' (*S. ferdinandi-coburgi* x *S. sancta*)

This is another old hybrid whose chief merit lies in its amenable constitution and ability to thrive on the rock garden. From the dark green foliage many 2 to 3 inch flowering stems are thrown up each of which bears four or five rather thin flowers. The petals, though small, are of the deep golden-yellow associated with the flowers of *S. ferdinandi-coburgi* and the overall effect is therefore quite rich.

S. 'IRIS PRICHARD' (*S.* x *godroniana* x *S. lilacina*)

This hybrid is reputed to have some 'Engleria' blood in its pedigree but nevertheless it has all the appearance of a typical Kabschia and is here treated as such. Its particular distinction lies in the unique colouring of the flowers—an attractive buff-apricot. It is also distinguished by being one of the earliest of the hybrids to flower, commencing at the end of January. The flowering stems are approximately 2 inches high and mostly bear three flowers. The foliage is attractive and distinct, the rosettes being markedly silvered with lime encrustations, individual leaves recurving outwards to terminate in a spiny tip.

S. 'IRVINGII' (*S. burseriana* x *S. lilacina*)

This is an old hybrid bred by Mr. W. Irving of the Royal Botanic Gardens, Kew. Its flowers are a very pale lilac-pink or blush in colour with the eye stained a deeper lilac-pink and are borne singly on one-inch flowering stems. The foliage is particularly distinctive, forming as it does a very neat, irregularly mounded carpet of closely set glaucous-grey rosettes which is always attractive even when the plant is not in flower. It is very free flowering but not quite so robust as *S.* 'Jenkinsae' which it resembles closely.

## S. 'JENKINSAE'
The flowers of this hybrid are almost identical with S. 'Irvingii' but the foliage is neither so neat nor so tightly set. It is also a more vigorous grower. The flowers, also borne singly, are carried on 2-inch stems. It is a reliable plant and a rapid grower, and should be in every collection. It will even thrive on the rock garden without extensive preparations. Its appearance suggests a similar parentage to S. 'Irvingii'.

## S. 'LADY BEATRIX STANLEY' (S. x godroniana x S. lilacina)
Like S. 'Iris Prichard' this hybrid is also reputed to have some 'Engleria' blood in its pedigree but it has nevertheless a typical Kabschia appearance. Its flowers are a rich red in colour and are very freely borne. It is rather slow to build up into a sizeable plant.

## S. 'MEGASAEFLORA'
This hybrid has handsome flowers of an unusual shade of rose-pink. The large petals are separated as in Megasea giving the flower a very distinct and readily recognizable 'clawed' make-up. The foliage is of the burseriana type.

## S. 'MYRA' (S. scardica x S. lilacina)
This is an old hybrid raised by Farrer and must have been one of the first to absorb the genetic possibilities of S. lilacina. The flowers are a deep pink. The whole plant resembles S. 'Cranbourne', except that the flowers are deeper in colour, and the constitution of S. 'Myra' is rather less easy-going.

## S. 'PETRASCHII' (S. marginata var. rocheliana x S. tombeanensis)
This is a reliable and striking hybrid, with fine grey-green hummocks of foliage and large well-formed white flowers opening from deep red buds. The flowers are borne on 2-inch long flowering stems, usually three to a stem. The contrast in colour between the open flowers and the buds is most attractive. It is one of the later flowering hybrids and is at its best in April

## S. 'VALERIE FINNIS' A.M. 1962 (illustration p. 19)
This is probably one of the two best yellow Kabschia hybrids. It forms a dense hummock of grey-green spiky burseriana type foliage, which is almost covered with large broad-petalled pale yellow flowers which are borne singly on short pink stems. It has proved to be a reliable grower with a good constitution. S. burseriana is obviously one of its parents. It was raised by Miss Valerie Finnis at Waterperry Horticultural School.

## S. 'WINIFRED' A.M. 1963 (illustration p. 20)
This is an outstanding hybrid resembling S. 'Cranbourne' in habit and foliage but with flowers of a rich crimson, deeper than any of the other Kabschia hybrids. The colour is deeper towards the

33

base of the petals. The large flowers are borne singly and sit almost directly on the foliage. One parent is said to be *S. lilacina*. It was raised by the late G. W. Gould of Nottingham and named after his daughter.

---

## PART 3. NEPHROPHYLLUM SECTION

WHILST IT CANNOT BE CLAIMED that the Nephrophyllum Section of the genus *Saxifraga* is an important one horticulturally, it would be wrong to write it off as having no horticultural significance. The Section contains a number of interesting and beautiful species which, although they may not be everybody's plants, are in many cases not difficult to grow with a little care and the protection of an alpine house during the winter. There is, in fact, one species, *S. granulata*, for which no special precautions are needed and which in its double-flowered form, under the name of 'Fair Maids of France,' is widely known and grown in gardens (Illustration p. 37).

It will, I feel, be helpful if we take a brief glance at Engler and Irmscher's grouping of the species within the Section, as it includes plants of very divergent natural distribution and habits. This classification splits them into four groups as follows:

### Arachnoideae

This group contains only one species, which is of little horticultural significance. It hails from Northern Italy where it is a cave-dweller found only in a very limited area.

### Irriguae

This group contains two species occurring in Spain and Asia Minor respectively, both of which have valuable horticultural potentialities.

### Granulatae

This group contains about a dozen species which, generally speaking, are plants of a Southern European distribution with Spain and Portugal as a particularly rich centre. *S. granulata* itself has already been referred to briefly and this species is exceptional in the extent of its range and variability, occurring as it does throughout the whole of Europe as well as in Asia and North America. Of the remainder only one or two have been introduced into cultivation and even these are very rarely seen. There is no doubt that the group contains some beautiful and fascinating plants and I believe that its horticultural potentialities have not yet been fully exploited.

## Sibiricae

This group contains about six species of Arctic and Sub-Arctic distribution which range through North America, Northern Europe and Siberia. They have very limited horticultural potentialities— at any rate in Great Britain.

Some, though by no means the majority of the species, suffer from that unfortunate affliction of biennialism which in the eyes of so many gardeners reduces their appeal. One should not, however, be put off by this.

As might be expected in a Section whose members come from such a wide range of habitats, a considerable diversity of physical characters is found. It is, in fact, difficult to put one's fingers on a hard and fast set of characteristics by which a species can be assigned with certainty to this section. Nephrophyllum derives from the Greek *nephros* (kidney) and *phyllon* (leaf), suggesting the possession of kidney-shaped leaves, but a number of species are included to whose leaves this description seems singularly inappropriate as they may be palmate to deeply divided or even, in a number of species, biternate. In almost all cases, however, the leaves are covered with hairs, usually glandular, and are particularly brittle to the touch, so that it is difficult not to break off good leaves at the same time as one is removing dead ones. These glandular hairs, when present, make the plant characteristically sticky to the touch. Another characteristic, which tends to run through this Section, is the presence of very long petioles to the leaves.

In the majority of cases the possession of bulbils growing either in the axils of the basal or the stem leaves is a characteristic trait of the section, these bulbils forming an efficient means of propagation and distribution. Here again there are exceptions without any bulbils.

In almost all cases the colour of the flowers is a clear white in which there is no trace of cream.

A further characteristic, again by no means unfailing, is that the majority of the species are deciduous and the foliage is renewed annually from the bulbils clustered at or below soil level, or scattered from the axils of the stem leaves.

On glancing through past numbers of the *Bulletin* seeking references to these plants, I have been struck by the fact that, with few exceptions, the references are to the finding of the plants in the wild and not to their cultivation in the garden or alpine house. A noteworthy exception is the reference in Vol. 3 (1935) to Dr. Roger Bevan being awarded Botanical and Cultural Certificates for showing a fine plant of *S. latepetiolata* earlier that year. It would seem therefore, that there is scope for experimenting with the cultivation of these plants. As their season of flowering is mostly June and July, they come at a time when there are few alpines in flower and when their interest and modest beauty can be all the more appreciated.

I cannot set up as an expert on this Section, for I have only grown five of the species and of these I have only seen two growing in the wild. However, I have been impressed by the species I have grown and consider that they are in some cases less well-known and distributed than they deserve to be.

Professor Webb, in the first volume of *Flora Europaea*, has as might be expected, dealt botanically with a greater number of species than are mentioned in the horticultural reference books, including the Royal Horticultural Society's *Dictionary of Gardening*. However, it is evident that a number of the species, particularly the Spanish ones, form a complex so that in some cases there is legitimate room for doubt as to their exact status. As gardeners, however, I do not feel that we need get involved in these aspects of a difficult group. In the main I shall base the nomenclature used in this article on that of *Flora Europaea*. The species on which I shall concentrate will be those which are in horticultural circulation, or which seem to justify being introduced or re-introduced. The remaining species, which are considered to be of purely botanical significance, will only be referred to briefly.

## Arachnoideae

### S. ARACHNOIDEA

Farrer in his *English Rock Garden* gives a typical and delightful description of this inhabitant of the limestone caves in the Cima Tombea area west of Lake Garda. It is a small annual or biennial of lax growth with oval, toothed leaves and small pale yellow flowers. I have seen it growing beneath the benches in the alpine house of the Munich Botanic Garden (See Herr Schacht's description in Vol. 32 p. 139) where it is kept as dry as in its native haunts. It is a plant more of botanical interest than garden value.

## Irriguae

### S. LATEPETIOLATA (Illustration opposite)

This saxifrage only occurs naturally in a limited area of Eastern Spain whence it was introduced to cultivation in this country by Dr. Giuseppi, though it had been known to botanists since 1874. Botanical and Cultural Certificates were issued to Dr. Roger Bevan when he showed it in London in 1935.

*S. latepetiolata* is a very definite biennial which dies off completely after flowering. In its first year of growth it makes a very symmetrical rosette of long-petioled, kidney-shaped leaves. These leaves are

*Saxifraga latepetiolata* (opposite)                    Photo: W. Harding

37

*Saxifraga granulata fl. pl.* (p. 39)                    Photo: R. Elliott

*Saxifraga irrigua* (opposite)

noticeably hairy and clammy to the touch. In the second year a handsome pyramidal inflorescence is thrown up which continues to open flowers and be very decorative for at least six weeks. This inflorescence is a foot high when well grown and nearly as wide at the base. When in flower the whole effect is still very symmetrical. In this country it flowers in late May and June.

I have never found that it has been affected by frost or snow, in spite of its Spanish origin. I have left a plant outside uncovered for most of the winter and only weakened and taken it under cover when the damp conditions were beginning to cause some of the leaves to rot. I think that, to have it looking at its best, it is almost certainly advisable to bring it under cover in the winter. I can recommend it as a plant well worth growing.

## S. IRRIGUA (Illustration opposite)

This saxifrage has its centre of distribution in Asia Minor. In Europe it is found only in the Crimea. It is said to be monocarpic and is probably best treated as a biennial, but with me plants have flowered well for a second time. The light green sticky leaves are deeply cut with some resemblance to buttercup leaves. No bulbils are borne in the axils of either the basal or the stem leaves. The flowering stems are about a foot in height and are much branched, the branching being in the upper half only. The fairly large white flowers are pleasant but not spectacular. This is one of the more easily grown species though the Section contains more beautiful plants. Nevertheless it could well be grown more frequently.

## Granulatae

### S. GRANULATA (Illustration p. 37: in its double form)

This is a plant with an extremely wide distribution occurring as it does through North America, Northern and Southern Europe and Asia. Not unexpectedly for a plant with such a wide range, it varies considerably, but there is only an incomplete link of forms over its range—as far as present knowledge goes.

Professor Webb has shown that there are differences between those plants found on granitic rocks and those found on calcareous strata and has consequently afforded them sub-specific rank. It seems very strange however to the horticulturist that in Northern Europe and particularly in this country, it is a plant that grows in moist lowland meadows on sandy and gravelly soils, whereas in Southern Europe it is said often to favour dry, rocky situations. When I have found it in Southern Europe, however, it has normally been growing on grassy slopes, often in shade for a portion of the day.

I feel that *S. granulata* is the one member of the Nephrophyllum Section which can be considered a thoroughly good-tempered and reliable outdoor plant in this country. In many gardens it becomes naturalized to give a very pretty effect. Surprisingly it is most often the double form which is offered by nurseries even though the single plant makes the more graceful addition to the alpine garden.

The leaves are mostly basal and the flower stem rises to a foot high, sparsely clothed with smaller leaves and bearing at the summit a loose cyme of large white flowers. The bulbils are all at the base of the stem underground.

## S. CORSICA

This species is very closely allied to *S. granulata*, but differs from it by the basal leaves being more or less 3-lobed and the stems branched low down to give a more diffuse inflorescence, about eight inches in height. *S. corsica* also bears basal bulbils. It occurs in Corsica, Sardinia, the Balearic Isles and East Spain. The Spanish form differs slightly from the Corsican form and is known as *S. corsica* ssp. *cossoniana*. (Syn. *S. cossoniana*).

This species is only just in cultivation in this country. I have grown ssp. *cossoniana* as an alpine house plant, and found it easy to grow and very beautiful. It caries elegant large white flowers: the foliage dies down soon after flowering.

## S. ATLANTICA, S. CARPETANA, S. DICHOTOMA and S. GRAECA

These four species are very closely allied to each other and in fact their standing as four separate species is somewhat doubtful. They are all of the *S. granulata* persuasion and are mainly centred in Spain although *S. graeca* occurs also in the Balkans and Southern Italy. The stem leaves are more numerous and incised and the inflorescence more compact than *S. granulata* but there would seem to be little to recommend the separate introduction and cultivation of these species.

## S. BULBIFERA

This is another species closely allied to *S. granulata*, but it is taller, up to 15 inches in height and bears bulbils in the axils of the stem leaves. Only three to four flowers are borne in a very compact cyme. It is a rather gaunt plant, not, so far as I know, now cultivated in this country and of no great horticultural merit. It grows in Central and Southern Europe.

## S. HAENSELERI

This species, although still allied to *S. granulata*, is very distinct in general appearance as it is much shorter, only 3 to 5 inches tall, with the flowers borne in a branching inflorescence from the neat tufts of basal leaves. The bulbils are underground. It is covered with glandular hairs and is very sticky to the touch. *S. haenseleri* comes from Southern Spain where it grows, amongst other places, in the cliffs of Grazalema. All eye-witness accounts, including that of Dwight Ripley (Vol. 12 p. 48), testify to it being a charming plant of fragile appearance.

It is not, so far as I am aware, in cultivation in this country and is a species which ought to be introduced in the future, even if it is not of the easiest cultivation.

S. BITERNATA (illustration p. 20)

We now come to another group of species which, although still botanically allied to *S. granulata*, are very distinct in appearance and are distinguished by the possession of leaves which are ternately lobed or divided. This characteristic is most marked in *S. biternata* where the leaf is divided into three quite separate stalked leaflets each of which is composed of three segments. The flowers are large, pure white and very handsome. It bears 'bud-like' bulbils in the axils of the leaves which, after flowering, grow out as young plant-lets still attached to the parent plant. The long-petioled leaves are particularly brittle to the touch and readily snap off. It is found only in a limited area in Southern Spain, notably at El Torcal de Antequera.

This is a most decorative plant and very well worth growing. It is in cultivation in this country in Botanic Gardens but only rarely seen outside. Although stated in the R.H.S. *Dictionary of Gardening* to be rather tender, I have not found it to show any ill effects from fairly severe freezing in a cold greenhouse or frame and have even over-wintered it in the open.

S. GEMMULOSA and S. BOISSIERI

These species are closely allied to *S. biternata* but are quite distinct from it. They have ternately divided leaves but a much laxer habit, with much branched inflorescences bearing a large number of relatively small flowers. Both species occur only in Southern Spain. From descriptions and illustrations there does not seem to be anything to recommend either species as garden plants. If they have been in cultivation in the past they appear now to have been lost.

Sibiricae

S. CERNUA

This is a rare native species occurring very locally in Scotland. It has an arctic and sub-arctic circumpolar distribution but occurs as far south as the Alps and Carpathians. It is about 6 inches high with bulbils in the axils of the stem leaves and a solitary terminal drooping flower (if it flowers at all). It needs moist, but not stagnant, shady conditions and is a most difficult plant to cultivate in this country, with no outstanding merits to justify the effort. Seed is however offered from time to time in the A.G.S. Seed List and one assumes that this is collected in North America, as it rarely if ever seeds here.

## S. RIVULARIS

This is another Arctic and Sub-Arctic species occurring very sparingly in Scotland. It seems to need brookside conditions and is difficult to cultivate. The habit is more trailing than that of *S. cernua.* and it bears more than one flower. The flowers are however small and of no great interest. Seed is offered occasionally in the A.G.S. Seed List and one assumes that this also is collected in North America or elsewhere. Its horticultural potentialities are very small.

## S. SIBIRICA

This is a small tufted species which hails from Eastern Russia, South-Eastern Bulgaria and North-Eastern Greece. It also grows in Northern and Central Asia. The flowers are fairly large and are borne at the end of the rather frail drooping stems. Farrer praises its 'delicate daintiness of charm' and says that 'it is as easy to grow as it is hard to come by'. It is not, so far as I am aware, in cultivation in this country at the present time and seems to be worthy of being introduced again.

## S. EXILIS, S. CARPATHICA and S. DEBILIS

These three species are close relations of *S. sibirica* and do not appear to have any great horticultural possibilities. They are not, so far as I am aware, in cultivation in this country.

*S. carpathica* is found in the Carpathians and South-Western Bulgaria.

---

# ROBERTSONIA SECTION

This is a Section of useful low evergreen plants characterized by the familiar 'London Pride'. None are very difficult to grow and all have the useful attribute of being able to flourish and flower effectively in quite deep shade. In addition, they are not particular as to soil and can be increased by simple division. Plants of such good temper and useful attributes are in great danger of being despised for their very ease of cultivation, so that some detailed information about this Section will not come amiss in a publication where references to them are few and far between. At the present time, when gardens are normally completely maintained by their owners, interest is being increasingly shown in ground-cover planting to eliminate bare soil and weed growth, and it is in this context that the members of the Section merit particular attention. In the choicer parts of the rock garden, where the conditions of light and drainage favour the more exacting and exciting species, we may not grudge the labour of hand weeding, but for that very reason we should be grateful that there are decorative plants which will solve our problems completely and painlessly in the less favoured areas.

Whilst this Section contains only a limited number of species, it is unfortunate that their inter-relationships are somewhat more complicated than is realized by most gardeners. We are used to seeing in our catalogues the familiar reference to *S. umbrosa*, 'London Pride'. Unfortunately the true *S. umbrosa* is not the plant which we all know and grow as 'London Pride'. It is with some diffidence that I mention this fact for I can already feel the curses of thousands of enraged gardeners falling upon my head. However, as I am treating this Section in some detail, I feel in honour bound to point out this and one or two other inconvenient facts which are now well-known to botanists.

## The Umbrosa Group

Botanically, the Umbrosa group consists of three species, SS. *umbrosa*, *hirsuta* and *spathularis*. Although generally acknowledged to be valid species, they are closely related and, when grown in proximity, they cross with each other to produce hybrids of some constancy and importance.

The three hybrids, which have been produced, are:—

S. x *urbium* (*S. umbrosa* x *S. spathularis*)
S x *geum* (*S. umbrosa* x *S. hirsuta*)
S. x *polita* (*S. hirsuta* x *S. spathularis*)

S. x *urbium*

Of these three the first is the true 'London Pride' which is unknown in the wild and most probably originated in cultivation, displacing its parents due to its very great merits as a garden plant. This is another case of 'mistaken identity' rather similar to that of the common Catmint, *Nepeta* x *faassenii* (syn. *N. mussinii*).

The second of the three hybrids, *S.* x *geum*, is the plant which was called for so many years *S. geum* and was thought to be a true species. Unlike *S.* x *urbium* ('London Pride') this cross is known in the wild, occurring with some regularity where the areas occupied by the two species overlap, as in the Pyrenees.

The third hybrid, *S.* x *polita*, occurs quite commonly in the wild but has not been considered sufficiently attractive to be introduced into cultivation.

With this preliminary explanation and with the cautionary word that the species in the Robertsoniana Section tend to vary considerably in their leaf characters, let us now look at each species and hybrid with an eye mainly for its actual or potential garden worth.

### S. X URBIUM. 'London Pride'. (*S. umbrosa* of gardens)

This cottage garden plant was certainly in cultivation in this country back in the 17th century and possibly even earlier. Consequently it features prominently in the fascinating 17th century garden behind Kew Palace recently constructed under the aegis of Sir George Taylor, the Director of the Royal Botanic Gardens. One has no need, however, to look thus far, as it is found in every typical cottage garden where its sterling merits are well recognised and appreciated.

In the days when Walter Irving wrote his book *Saxifrages or Rockfoils*, a large number of varieties and cultivars were distinguished by name. Most of these now seem to have vanished, apart from a golden variegated cultivar, 'Variegata Aurea', a smaller serrated-leaved cultivar 'Colvillei' and the small-leaved and dainty 'Primuloides' cultivars. The last were not mentioned by Irving unless they were embraced in his variety *minor*.

At the present time the 'Primuloides' cultivars are probably of more interest to Members than the typical 'London Pride' and its other cultivars. Their neatness of habit and daintiness of bloom make them less invasive and more aristocratic than the latter which are, perhaps, best relegated for use as a ground cover in shrub borders or the forefront of general purpose borders. The original *primuloides* variety (*S.* x *urbium* var. *primuloides*) is reputed to occur naturally in the Pyrenees, where it bears small leaves and pale pink flowers on 6 to 8 inch high flower stems.

The cultivar 'Elliott's Variety' or 'Clarence Elliott', as I suppose it should now more properly be called, is a particularly fine form originally selected by the late Clarence Elliott and one of the finest plants which he put on the market. This is even smaller and neater than var. *primuloides* and has red stems 5 to 6 inches high bearing an inflorescence of very pretty little rose-pink flowers. It is a plant well worth growing in any rock garden and can be seen nowadays in most collections. It prefers some shade and flowers well on a starvation diet. The cultivar 'Ingwersen's Variety', more properly 'Walter Ingwersen', is said to be an even smaller plant but I have never been able to see any difference between them.

'Variegata Aurea' is a golden-variegated cultivar seen at its best when growing in full sun. Its leaves are blotched with gold but I cannot say that I find it very attractive even though it is currently being shown by a number of nurserymen.

'Colvillei' has serrated leaves and white flowers but is of no great merit.


### S. UMBROSA

This, the true *S. umbrosa* as described by Linnaeus, does not, so far as is known, appear to be in cultivation in this country, having over the years been displaced by its more vigorous hybrid progeny described above. It is distinguished from the latter by having the petioles slightly longer than the leaf blades with conspicuous cartilaginous margins to the leaf blades. The leaves are larger and less spreading. It is of no particular horticultural value but one would be interested to know whether the species still lingers on in gardens in out of the way places.

*S. umbrosa* is a native of the West and Central Pyrenees and does not occur as an indigenous plant in the British Isles as do *S. hirsuta* and *S. spathularis*. It is, however, naturalized at Heseldon and Linn Gills in Yorkshire where it has been known since 1792.


### S. HIRSUTA

Unlike the last species, *S. hirsuta* is a true native of Ireland, though more plentifully distributed in the Pyrenees. It is like a hairy oval-leaved 'London Pride' but requires damp conditions with a rather humid atmosphere to be really at home. It has no special merit as a garden plant except in such conditions.

### S. SPATHULARIS

This is another Pyrenean species which also spreads up the Western Coast of Ireland but, except as one of the parents of 'London Pride', it has no very great horticultural value. It is occasionally offered in the nursery trade.

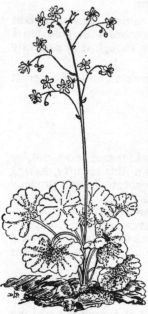

### S. X GEUM

This plant has, until recent times, been considered as a true species, *S. geum*. However, the fact that it is now known to be a hybrid neither decreases nor increases its horticultural significance. It seems to have been grown considerably more in the past than it is now. Though not a native of Great Britain, like *SS. hirsuta* and *spathularis*, it is a plant which occasionally naturalizes itself as a garden escape and Walter Ingwersen wrote in Volume 2 of the Bulletin of one particularly well-known colony in an open cave between Ingleton and Ribblehead in Yorkshire. It is not quite such an easy plant to grow as *S. x urbium*, at any rate in the south, for it prefers a rather moister climate, but it is certainly not difficult if suitably placed and cared for. The leaves, with petioles several times longer than the blades, have a certain elegance which is lacking in the grosser forms of *S. x urbium*.
It is, in the wild, a plant of very variable leaf form and in its heyday a number of its more extreme forms were selected and sold as varieties under such names as *crenata, dentata, elegans* etc. The fact that these varieties do not appear to be sold any longer is no great loss. It is, however, a pity that *S. x geum* itself has become so comparatively scarce and is at the present time obtainable from so few nurseries. If people will ask for it, and it is a plant worth growing as an attractive ground-cover in shady places, it will no doubt become readily available again.

*Saxifraga valdensis* (p. 57)                    *Photo: D. F. Merrett*

47

*Saxifraga exarata* (p. 91)                        *Photo: R. Elliott*

*Saxifraga cotyledon* at Macugnaga (p. 64)          *Photo: D. Holford*

*Saxifraga* 'Tumbling Waters' (p. 70)          *Photo: Roy Elliott, A.R.P.S.*

*Saxifraga aizoon* (*paniculata*) (p. 58)          *Photo: M. G. Hodgman*

# Other Species and Hybrids

### S. CUNEIFOLIA

This species is widely distributed in the wooded sub-alpine mountains of Southern Europe, extending from the Pyrenees to the Carpathians, and has been in cultivation in this country for almost two hundred years.

In the wild it occurs in two distinct forms (apart from the species itself), varieties *capillipes* and *subintegra*. Var. *capillipes* is a smaller form with barely scalloped spoon-shaped leaves and flowers speckled with purple whilst var. *subintegra* has somewhat larger bluntly toothed leaves, and flowers with two instead of one yellow spot on each petal, as does the species itself.

*S. cuneifolia* and its varieties are pretty little carpeters which will cover the ground in any odd corner of the rock garden, in broken sun as well as in shade. In May and June they will throw up a large number of 6 inch high panicles of flowers, more or less white in colour, with coloured markings. This is a species that should be grown more widely, both for its daintiness and for its qualities as a ground coverer.

### S. X ANDREWSII

Before concluding this section, one inter-sectional hybrid of some importance will be referred to. This plant, *S. x andrewsii*, is usually considered to have originated as a cross between *S. hirsuta* and *S. paniculata* (syn. *S. aizoon*), although either *S. umbrosa* or *S. x geum* is sometimes given as one of the parents. That, however, is probably due to the nomenclatural confusion which has existed in the past in this section. As it was collected from the wild at the end of Glen Caragh in Killarney, Eire, it seems certain that one of its parents must have been a local wild plant, even if it was fertilized by pollen from a garden plant of *S. paniculata*. This points to *S. hirsuta*, a true native fairly widely distributed in Western Ireland.

Whatever its origin, *S. x andrewsii* is a plant of really sturdy constitution and sterling merits as a garden plant. It will grow in sun or shade, but makes a better plant in partial shade. It is also not of so rapidly spreading and invasive a habit as *S. x urbium*. It bears an inflorescence about 10 inches high and rather similar to the latter except that the individual flowers are twice as large and not quite so numerous. The pink ovary and the red spotting of the petals give the flowers an overall flesh-pink tinge which is attractive. The flower stems are distinctly red. The leaves are long and narrow and sharply toothed along their margins. The rosettes of leaves are tightly clustered together to give a compact plant. It is interesting to note that this hybrid is widely grown in the British War Cemeteries

51

in France where the gardeners, from their long experience, have found it to be a useful, easily-propagated and distinctly decorative plant.

This hybrid deserves to be more widely known and grown, as does *S.* x *wildiana* which is very similar.

---

## EUAIZOONIA SECTION

### (The Encrusted Saxifrages)

AFTER THE FAMILIAR LONDON PRIDE and the ubiquitous Mossy Saxifrages the Encrusted Saxifrages are the next most commonly grown in gardens, though I gain the impression that they are now less popular than they were in the past. With their almost universal ease of cultivation, their attractive evergreen foliage and their often strikingly beautiful sprays of flowers it is not easy to understand why this should be the case. One can only presume that it has come about as a result of a change of horticultural fashion, aided perhaps by the present high cost of stonework which reduces the opportunities for the construction of those rock crevices in which these plants grow and look so well. However, although their hard lime-encrusted cushions never look better than when grown in this manner, with their arching sprays of flower falling outward into space, they do not in fact demand such conditions for their healthy survival and most can be planted on the flat, providing the drainage is good. Even though many are so good-tempered that they will produce good results with the minimum of preparation, it nevertheless pays to incorporate chippings and humus into the area where they will be planted in order to sharpen up the drainage and to provide a suitable root-run for the years they will occupy the same position. This will also ensure more handsome rosettes of foliage and finer panicles of bloom.

This Section has in the past suffered from the selection and naming of too many insufficiently distinct varieties and hybrids, which arise from the polymorphic nature of some of the species and the extreme facility with which most will cross, when grown in proximity with each other. The Section is one that is ill-suited to stand such treatment as in the majority of cases, unlike the Kabschias, the colour of the flowers is white or cream, sometimes with faint red spotting on the petals, so that varietal distinctions have often been difficult to describe and observe and are of no great horticultural significance.

When in 1951 *Saxifraga* 'Southside Seedling' appeared, with the red spotting run together in large areas so that the flowers appeared to have an equal patterning of red and white, the gardening public were at once struck by this spectacular break in a Section from which they had ceased to expect any striking novelties.

As true breeding from seed is so difficult to ensure with these species, it is indeed fortunate that they can, with few exceptions, be propagated vegetatively with ease. In most cases every rosette which is detached will readily root in any moist open medium. Many species can in fact be increased by simple division. In one or two cases however, where the species do not readily make offsets and die after flowering (i.e. monocarpic), it is essential to resort to seed, taking all possible precautions to prevent crossing with another species.

When approaching the task of writing up this Section I remembered reading a most comprehensive treatment of the subject by Dr. R. C. C. Clay which appeared in an earlier *Bulletin* (Vol. 15, p. 87). On re-reading this I was again struck by the detailed knowledge and experience of the author. How was I to improve on this? My first instinct was to omit this Section and simply to refer readers to this earlier article. On second thoughts, however, I realized the undesirability of this course of action for few present day readers can be expected to possess the *Bulletin* of twenty years ago. Also the passage of the years has inevitably brought changes of emphasis and nomenclature, together with the introduction of a few new and worth-while hybrids and the eclipse and disappearance from cultivation of others. This has made an up-to-date assessment essential.

I must also admit that my approach is not quite the same as Dr. Clay's, for it is my intention to simplify the picture, even sometimes at the risk of oversimplification. Dr. Clay set out to record faithfully the whole recognized range of varieties, hybrids, and cultivars to which names had been given, whereas I shall omit reference to all those which have in my opinion no valid claim to separate cultivation and perpetuation. Even from the point of view of the pure botanist, the desirability of ascribing names to every small variation in such a polymorphic species as *S. aizoon* is now not admitted. It must be accepted that a number of the species in this Section are extremely variable, particularly in relation to such characters as the length and width of their leaves and the extent of their silvery encrustations.

### Crustatae

S. CALLOSA (illustration p. 60)

This is a variable species with narrow untoothed leaves about 2 to 3 inches long and spatulate at their extremities. It occurs in Spain, Southern France and Italy. The forms found in Italy and the

Maritime Alps are nowadays grouped under subspecies *callosa*, whereas the rather more homogeneous populations found in South-Western France and North-Eastern Spain are grouped under subspecies *catalaunica*. The latter was, in the past, treated by some authors as a separate species, *S. catalaunica*, on account of the broader leaves and the glandular-hairy and wider spreading inflorescence. No. doubt, however, there are sound botanical reasons for treating them all as forms of the same species and so we should adhere to the more modern treatment. It is unfortunate that these plants are still often met with in gardening books and catalogues under the epithet of *S. lingulata*. This name, though admirably descriptive, must now be buried in the past.

A good case could be made for calling this the most generally graceful and beautiful of all the Encrusted Saxifrages and certainly, when one sees it well grown, one wonders why so many of the hybrids have commanded such popularity and have continued to be propagated and sold. It may not be such a spectacular plant as *S. longifolia*, as seen in its native Pyrenean haunts, but at least one does not have to wait many years for it to flower only to see it die after the brief consummation of its glory.

The silvery leaves of *S. callosa*, when grown healthily, are always beautiful, and from the rather irregular and spidery rosettes, the wand-like inflorescences, composed of numerous pure white flowers, closely grouped along the upper half of the flowering stems, arch outwards with great elegance. In many of its forms these panicles are all the more attractive for the one-sided arrangement of the flowers in the inflorescences. The rosettes usually build up into an irregularly humped mass, only the largest throwing flowering stems in early summer, while the smaller rosettes continue to build up their strength in preparation for flowering in a subsequent year.

The temptation is to grow these plants, which look so hard and sun-defying, in full exposure to the sun and this temptation is not diminished by the fact that, unlike the Kabschias, they will probably withstand baking without a spectacular demise ensuing. It will, however, always be wise to plant them where they get shade for a portion of the day, as in such a position they will grow more healthily and throw out more handsome panicles of flower. In nature they favour the shadier positions and, even though we do not in these islands receive the full ardour of the Mediterranean sun, they will appreciate some shade, at any rate in the south.

Different forms are advertised in catalogues under the names of var. *albertiana*, var. *bellardii*, var. *lantoscana* and var. *australis* and there is no doubt that there are differences between the selected forms that are propagated under these names. It is less clear whether these forms, as distributed in the trade, are all selections of the pure species or whether some have an admixture of blood from other species. The form which is my favourite and which I have met in a number of places in the Abruzzi mountains is the form which Engler and Irmscher placed under var. *australis*. This is the longer-leaved form which occupies the southern part of the range and occurs in the Central and Southern Apennines, Sardinia and Sicily. This form, grown from seed, has flowered so magnificently in my garden that there seems little point in continuing to foster so many variants of the species. It is illustrated on page 60 where its 12 to 15 inch long wands of flowers can be seen to good advantage.

Of the other named varieties, var. *lantoscana* is, as found in the wild, a well-defined form which is somewhat smaller in all its parts with shorter leaves which are broader and more rounded at the ends. A nurseryman's type, *lantoscana* 'Superba', is similar but with a longer inflorescence. Var. *bellardii* is synonymous with sub-species *callosa* in its typical northern form, var. *australis* being synonymous with sub-species *callosa* in its typical southern form.

The hybrids of *S. callosa* with other species, whether natural or man-made, are dealt with under the section devoted to Encrusted Hybrids.

## S. COCHLEARIS (illustrated on p. 60 in the 'Minor' form)

This species varies in size in all its parts from one locality to another, but otherwise its characteristics are reasonably constant and readily recognisable. It comes from a fairly restricted area, being confined to the Maritime Alps on either side of the French-Italian border. It is always a neat plant and has short, untoothed, heavily lime-encrusted leaves, each of which is sharply expanded towards its extremity into a thickened shell-shaped termination, which is quite distinct from that of any other species apart from some resemblance to *S. valdensis*. The latter can, however, be distinguished readily by characters which will be given under the latter species. The leaves vary in length from ⅓ of an inch to an inch and the regularly formed rosettes vary in diameter from ¼ an inch to an inch. The crowded rosettes are built up on each other to produce a markedly mounded plant which may even be higher than it is broad, or which, if constricted between adjoining rocks, may spread along the crevices in a very pretty manner.

The inflorescences, rising to a height of 6 to 9 inches, have a particularly light and airy character, due to the vividly white flowers being borne on slender, distinctly red, flowering stems which branch in the upper half only.

Most authors distinguish the species and varieties 'Major' and 'Minor'. I am not, however, convinced by this division and feel that we should more properly refer only to the type species and variety 'Minor'. The form which passes as 'Major' is—in my view-probably not a true species but has some *S. callosa* blood in it, since there are a number of significant differences between it and the type species other than size. I have not had the opportunity to study the so-called variety 'Major' in the wild, but plants received from a botanical garden source do not fit well into the specific concept of *S. cochlearis*. This opinion is advanced only tentatively. *S. callosa* does, however, occur more or less alongside *S. cochlearis* throughout most of its range.

*Saxifraga cochlearis 'Minor'*

S. C. VARIETY 'MINOR'. This is an extremely dwarf and compressed form of the species, differing in nothing except the size of its parts. It is a most desirable plant and should be in every collection of saxifrages, however small. The inflorescence rises from the diminutive $\frac{1}{2}$ inch diameter rosettes to a height of 4 to 5 inches. It was given the Award of Merit in 1964.

In nature and in cultivation *S. cochlearis* has crossed with most of the other Encrusted species in whose proximity it has grown, and has produced some worth-while hybrids which will be dealt with later on.

*Saxifraga crustata*

S. CRUSTATA

This species comes from the Eastern Alps and extends southwards into Jugoslavia—it is thus found in Italy, Austria and Jugoslavia, the Dolomites providing some of its commonest localities. It is not a very impressive species nor particularly interesting horticulturally, as its red-stemmed inflorescences bear their somewhat dingy yellowish-white flowers rather sparsely and are only about 5 to 9 inches high. The rosettes, which are crowded closely together into a mat, are however attractive in a restrained way with their narrow, silver-beaded, untoothed leaves which are about 1 to $1\frac{1}{2}$ inches long. This species grows best in partial shade.

It crosses in nature with *S. hostii* to give a hybrid which has been given the name of *S. x paradoxa*. This is described later on.

S. VALDENSIS (illustration p. 47)

This species occurs only in the Western-South Alps astride the French-Italian border, where it is restricted to the higher altitudes between 2,000 and 2,500 metres. Two of its classic localities are on Mt. Cenis and Mt. Viso.

It is a very tight little plant, found wedged securely into the crevices of the rocks, resembling a very compact form of *S. cochlearis* 'Minor' with which it is often confused. On close inspection, however, the individual leaves will be seen to be markedly different, as they do not broaden out into a semi-orbicular shell shape at their extremities, but rather into a thickened wedge with an obtuse point. The leaves are glaucous, lime-encrusted and reflexed.

The inflorescences are 2 to 5 inches high, with reddish densely-glandular stems, bearing 5 to 10 white flowers in a head that is distinctly more clustered than that of *S. cochlearis*. A very good idea of the general appearance of the plant in flower can be gained from the photograph on page 47 though my own plants have not given quite such a generous display.

This is a very slow growing and rather particular species, which is the least suitable of the Encrusted saxifrages, apart from *S. florulenta*, for outdoor culture in this country. As a pan plant, however, it is both fascinating and rewarding, its slow growth being meritorious under these conditions. The flowers are normally a good clear white and the inflorescence distinctly decorative. It flowers in May and June and appreciates full sun.

Some authorities have classified it as a Kabschia, rather than as a member of the Euaizoonia Section. Certainly its superficial appearance might tempt one to this conclusion, but the fact remains that it crosses readily with other Encrusted species such as *S. cochlearis* and has so far not been crossed with a Kabschia. This, to my mind, provides a strong prima facie case for retaining it in the Euaizoonia section.

S. LONGIFOLIA

This species occurs over a comparatively restricted range in nature, being found only in the Pyrenees and certain mountainous areas of Eastern Spain. It is found on both the French and the Spanish sides of the border. Plants in the wild vary somewhat in the width of the leaves, but in general it is, due to its narrower distribution, rather less variable than *S. callosa*. Both in its rosette and in the magnificence of its inflorescence, it is a readily recognised and very distinctive species. As mentioned previously it is monocarpic and the plant dies after its final riotous blossoming, which occurs only after it has built up its strength for a number of years.

The rosettes are extremely regular in formation and up to 6 inches or more in diameter. They are composed of very many narrow leaves, up to four inches in length, untoothed, smooth and lime-encrusted. The leaves are expanded in width towards their extremities. These rosettes are objects of great beauty in the years before they put forth their giant inflorescences, especially if growing from a shady crevice in a vertical wall of rock. Such a position, which it most often favours in nature, will allow it to show off its 2 foot long compact pyramidal panicles of bloom to the best advantage. These inflorescences are typically distinguished by having lateral flowering branches arising from them right down to the base, instead of only in the upper half as in other species. The petals are usually pure white, though rarely some red spotting may occur. The flowering stem and pedicels are glandular-hairy.

This species certainly crosses with *S. callosa* (ssp. *catalaunica*) in nature, as well as with a number of other species in cultivation. These hybrids will be dealt with in due course, but it is pertinent to remark here that the hybrids with *S. callosa* such as 'Tumbling Waters' and 'Walpole's Variety' have a particularly useful niche in the saxifrage lover's repertoire; they have eliminated the inconvenient effects of the monocarpic habit, by having the power to produce offsets with which future years' flowering display is guaranteed, without having to resort to the rather long and tedious procedure of growing the plant once again from seed.

**Peraizoonia**

S. AIZOON (*S. paniculata*) (illustration p. 50)

I have noted with some concern that in Vol. I *Flora Europaea, S. aizoon* has now been renamed *S. paniculata*, but I am afraid that, as a horticulturist, I cannot be converted to the use of this new, or rather very old, name until I can feel assured that it is here to stay and that the whole botanical fraternity are permanently converted to it. (There will be few gardeners who do not share these sentiments, but *S. paniculata IS* the correct name Ed.)

*S. aizoon* is a species with a very wide distribution throughout Southern and Central Europe, and extending into the Arctic Regions including Baffin Land, Labrador, Ontario, Quebec, Greenland, Iceland, the Faroes and Norway. As might be expected with such a far-ranging species, it varies considerably throughout its range and many of its local populations have distinctive characteristics. Plants of sharply contrasting forms can, however, be found in close geographical proximity so that its variation cannot be said to follow a tidy or logical pattern. To only one of its forms does Professor Webb give subspecific rank and that is the ssp. *cartilaginea*, which occurs in the Caucasus. Other authors sometimes treat this subspecies as a species in its own right.

The rosettes, which are almost always distinguished by markedly incurving leaves, with distinct serrations along their margins, vary very greatly in size from one form to another. The lengths of the

*Saxifraga florulenta* (p. 66)  *Photo: Roy Elliott, A.R.P.S.*

*Saxifraga callosa* (p. 53)

*Photos: W. Harding*

*Saxifraga cochlearis* 'Minor' (p. 55)

leaves may range from as little as 5 mm. to as much as 60 mm. in the extreme cases. Similarly, the length of the flower-stems varies from 4 to 30 cms. It must be clear from this that the extreme forms, if seen side by side, would hardly be recognisable as belonging to the same species. Furthermore, the degree of encrustation varies greatly from very intense in the dwarfest forms, to almost lacking in many of the larger green-leaved forms.

On the whole, I would not rate the different forms of *S. aizoon* very highly from the floral point of view as the majority have flowers of a rather undistinguished creamy-white, though sometimes the very profusion of flowering stems gives the effect of a foam of flower. Most of the flowers occur towards the summit of the stem, which is unbranched for much of its length. There are, however, two forms, *rosea* and *lutea*, with rose-red and yellow flowers respectively, which do achieve more floral effect, particularly when placed in proximity to each other.

As foliage plants throughout the year, I would rate a number of the forms very highly, particularly the more diminutive forms which make such tight, neat mats of lime-encrusted foliage, this development finding its most extreme form in *S. a.* var. *minutifolia* (syn. *baldensis*), from the vicinity of Lake Garda. The dainty dwarf forms are seen to the best advantage when grown in pots and pans where they are nearer to the level of the eye. The majority of the varieties, however, can be used almost anywhere in the rock garden, although some do better in partial shade and others in full sun.

Clay, in 1947, described thirty distinct varieties or subspecies of *S. aizoon*. A number of these forms are now quite unknown in commerce or the botanic gardens, but, in spite of this, I was surprised to find that a total of thirty so-called different forms are still listed in aggregate in recent plant or seed lists. If, however, all these forms were collected together I feel sure that many would prove to be indistinguishable. I do not wish to perpetuate this confusion, so I intend to describe only a dozen forms, all of which are quite distinct, all can be obtained from the nursery trade, and all have their separate merits justifying inclusion in a comprehensive collection of Encrusted Saxifrages. This, to my mind, represents the outside number worthy of separate cultivation, and includes the subspecies *kolenatiana*. The form *pectinata* is treated under the hybrids

### Saxifraga aizoon 'Balcana'

This form, said to be of garden origin, has fairly small, rather flat rosettes and white flowers heavily spotted with red. The flowering stems are stout and about 6 inches in height.

### Saxifraga aizoon ssp. cartilaginea

This subspecies, sometimes raised to specific rank, represents the eastern form occurring in the Caucasus and in the mountains of Asia Minor. The leaves are distinguished from the normal *S.*

61

*aizoon* varieties by ending in a fine point. The flowers are starry-white and are borne on flowering stems about 6 inches in height.

*Saxifraga aizoon* ssp. *cartilaginea* var. *kolenatiana*

This is a form of the subspecies *cartilaginea*, with long narrow leaves coming to a fine point and rose coloured flowers. It is said to appreciate partial shade. It is not now seen very often. It is sometimes listed as *S. kolenatiana*.

*Saxifraga aizoon labradorica*

This North American form is one of the more compact types, with rosettes only about ½ to ¾ inch in diameter and inflorescences about 3 inches in height. The leaves are well silvered and very decorative.

*Saxifraga aizoon lutea*

Although said by some authors to be of garden origin, this form is just as likely to have been selected from the wild, where the species can be found flowering in all shades of cream and pale yellow. This is a useful form which looks particularly well when planted alongside *S. aizoon rosea*. The flowering stems are 6 to 9 inches in height and the flowers are pale yellow.

*Saxifraga aizoon* 'Malbyi' (*Syn* '*Major*')

This is a very vigorous form, found in the mountains of Dalmatia, with inflorescences 12 to 18 inches in height. The flowering stems are red and the large rosettes are also often stained with red. The white flowers are heavily speckled with red.

*Saxifraga aizoon* 'Minor'

This is a miniature form, with small very silvery rosettes and flowering stems about 3 inches in height. The flowers are creamy-white.

*Saxifraga aizoon* 'Minutifolia' (syn. *baldensis*)

This is the most compact form, which hails from the mountains beside Lake Garda, notably Monte Baldo. The rosettes are barely ½ inch in diameter, and are very tightly packed together. They are heavily lime-encrusted and build up into a rather dome-shaped plant. The inflorescences, which are borne sparingly, are only 2 inches in height and are composed of small starry, creamy-white flowers. This is an admirable plant for crevice planting.

*Saxifraga aizoon orientalis*

This is the Balkan form of *S. aizoon* and is somewhat variable. These varying forms make a connecting link with var. *cartilaginea* of Asia Minor and its pink-flowered variety *kolentiana*. The medium sized rosettes are composed of very pointed leaves and give rise to inflorescences about 8 to 9 inches in height bearing creamy-white flowers lightly spotted with red.

*Saxifraga aizoon* 'Rex'

This is a good very silvery-leaved form with 8 inch high mahogany-red flower stems. The large round creamy-white flowers are unspotted. This is one of the handsomest varieties.

*Saxifraga aizoon rosea* A.M. (1907)

This form with rose-red flowers is said to come from Bulgaria. It would seem to be a natural variety rather than a hybrid, as it comes reasonably constant from seed. It is one of the prettiest of the *S. aizoon* varieties and well worth its place in any rock garden. The rosettes are large, yellowish green and about 1½ inches across and the flower stems are 8 to 9 inches high.

*Saxifraga aizoon venetia*

This is another very miniature form, hailing from the Venetian Alps. It makes rather a flat mat, unlike the humped plant of *S. aizoon minutifolia*. The flower stems are only 2 inches high and the flowers are pure white.


## S. HOSTII

The true species is nowadays relatively seldom seen in gardens and, when seen, it is usually in the variety *altissima*. It is however, one of the parents of a number of naturally-occurring hybrids, such as *S.* x *churchillii* and *S.* x *paradoxa* as well as several garden hybrids.

*S. hostii* is a species of the Eastern and Southern calcareous Alps in Italy, Austria and Jugoslavia, where it is not infrequent, though not occurring with the same ubiquitousness as its near relative *S. aizoon*. It is readily distinguished from the latter· by the blunt, strap-shaped, outwardly curving leaves, the rounded marginal toothing and the fact that it is larger in all its parts. It characteristically forms a broad flat mat of 3 in. diameter rosettes, which are dark grey-green in colour and marginally lime-encrusted. The the leaves are narrow, blunt-edged 1½—3 in. long.

The flowering stems in the typical form of the species are stout and erect and about a foot high, unbranched for the greater part and topped by a rather flattened, branching panicle of flowers of good form and substance. Each branch carries 3 to 5 creamy-white flowers which are often, though not always, spotted with red.

Professor Webb, in Vol. 1 *Flora Europaea*, distinguishes two subspecies, ssp. *hostii* occurring over most of the range, with the typically blunt-ended leaves, and ssp. *rhaetica*, with narrow leaves ending in a more acute apex, and only occurring in the Italian Alps. Of the subspecies *hostii*, a naturally occurring variety, *altissima*, is distinguished by the larger size of all its parts, with flowering stems up to 1½ feet high and leaves up to 4 inches long—this is the form which is most frequently seen in cultivation.

This species, which is distinctly decorative, could and ahould be grown more frequently in gardens, as it has a personality of its own. The mat-forming habit is effective and the erect flowering stems are more attractive than those of *S. aizoon*. Its culture is of the easiest.

**Cotyledoniae**

S. COTYLEDON (illustration p. 48)

Like *S. aizoon*, this is another widely distributed species occurring in the Alps, the Pyrenees, Central Europe, Scandinavia, Iceland and North America. Normally it seems to favour the granitic formations rather than the limestones. It varies considerably, both locally within any one area, and also from one region to another. These regional forms have sometimes been treated as sub-species, but here they will only be considered as varieties.

The rosettes of this species are rather flat with very typically broad strap-shaped leaves. The leaves in the different varieties vary considerably in size and in the ratio of breadth to length. The rosettes grow correspondingly larger if all lateral rosettes are suppressed. The glaucous leaves are distinctly toothed and lime-encrusted along their margins.

This species throws magnificent pyramidal inflorescences in June, when the rosettes, usually after two years, develop to flowering size. They are up to 2 or even 3 feet in length, and these spectacular plumes of flower are so striking that the plants require to be placed rather carefully, so that the flowers can be seen to the best advantage. Most often in nature, the rosettes grow out of the vertical rock, so that the massive inflorescences can arch out into space in the same way as *S. longifolia*. Of course, most of us cannot give it such a placing in our gardens, but nevertheless we can see that it is elevated as much as possible, and given space for its inflorescences to arch out without becoming an embarrassment. It can also be grown as a very effective specimen plant in a pot, and in this case all side rosettes should be removed as they appear, so that one large rosette 9 to 12 inches in diameter can be built up to give a particularly fine inflorescence. Anyone who has seen the plant used in this way in the Alpine Houses at Kew or Wisley will appreciate how effective it can be.

The flowers are typically white, with a green centre and un-spotted petals, but in a number of its varieties they may have either light or pronounced red spotting. In *S*. 'Southside Seedling'. which is often shown in catalogues as a variety of *S. cotyledon*, the red on the petals takes the form of a solid band. The parentage of this variety is, however, obscure and it is proposed to treat it under the hybrids.

The botanist recognizes two broad types amongst the forms and varieties of this species. There is a northern type, which Engler and Irmscher call form '*genuina*' and a southern type, which they call form '*pyramidalis*'. The northern form is distinguished by the possession of long broad leaves and an inflorescence which does not branch right from the base of the stem, whereas form *'pyramidalis'* is distinguished by shorter leaves, which are also not so broad, and by a very broadly pyramidal form of inflorescence, which branches from the base of the stem. They also refer to a form *'pauciflora'* which is found in Lapland, and is more of botanical than horticultural interest in that it bears inflorescences only 3 to 4 inches in height, bearing but few flowers. With respect, however, I feel that this sub-division is one that is not wholly reliable, as *S. cotyledon* is such a variable species wherever it occurs. I will, however, single out certained named 'horticultural' varieties whose characters have been fixed, even if their origin is in some cases obscure.

## *S. cotyledon* 'Caterhamensis' A.M.

This variety, which is reputed to be a selection from the northern type found in Norway, is one with the flowers heavily spotted with red. It is handsome and as easily grown as the type. This variety was given the Award of Merit in 1923, and the citation then referred to it as a very attractive hybrid. It is difficult to know whether the term hybrid was used loosely.

## *S. cotyledon* 'Montavonensis'

This variety occurs naturally in Central Europe and is only a dwarfer form of the species, with inflorescences about a foot high. It can be found growing in company with the species. It is not of any great horticultural interest, being inferior to the type. It was once wrongly given specific rank.

## *S. cotyledon* 'Norvegica' A.M.

This again is probably a naturally occurring form, which has been selected from the wild. It is certainly a northern form, though probably occurring elsewhere than in Norway, for instance in Iceland, Sweden and Finland. Though *S. cotyledon* 'Icelandica' is said to be a particularly fine form, I query whether it differs in substance from *S. cotyledon* 'Norvegica'.

The leaves are spatulate and about five eights of an inch broad, and differ from the type by tapering sharply to a short point. The flowering stem starts to branch from about 3 inches up from the base, and bears up to 50 side shoots, each carrying up to 20 starry, pure white flowers. It gained the Award of Merit in 1960.

## *S.cotyledon* 'Pyramidalis' A.G.M. 1925

This might perhaps be called more than a horticultural variety, as it agrees well with Engler and Irmscher's description of their form *'pyramidalis'*, which occurs locally in the Pyrenees and sparingly in the Valais and the Savoy Alps. It branches right from the base of the flowering stem, and makes a very broad pyramidal inflorescence, not quite as long as in some varieties but extremely handsome. The basal leaves are long, and narrower than in the typical form. It is a most useful early summer-flowering plant in the rock garden.

## Florulentae

### S. FLORULENTA (illustration p. 59)

This is another very local species growing only in the central area of the Maritime Alps astride the French-Italian border. It is a positive lime-hater and only occurs on the shady faces of the granitic formations, at altitudes from 2000 metres to 3250 metres. It is, without doubt, the rarest of the Encrusted Saxifrages in nature, as its area of distribution only extends for about twenty miles between Tende and Argentera.

This species is quite distinct in appearance from any other Encrusted Saxifrage, and this distinctness, combined with its very limited distribution, seems to suggest some relict species from ancient times, desperately hanging on in its final refuges.

It produces a flat rosette of very regularly arranged leaves, shown to perfection in our illustration, though in older specimens, due to the persistence of the dead leaves, the structure develops into an elongated cylinder with a flat green rosette at the summit. The leaves are dark green, spiny, pointed at the tip, and 2 to 3 inches long. They are without lime-encrustations. The rosettes, which are 4 to 6 inches in diameter, are slow to attain their full development and then, being monocarpic, die after flowering. The inflorescence, when finally produced, is, unlike *S. longifolia*, not very ornamental, the flowers being a rather muddy flesh colour. The latter are borne in compact narrow panicles about 18 inches long, but only very rarely in cultivation, the plants usually succumbing for one reason or another long before they reach this stage.

This is not at all an easy plant to grow, as it bitterly resents standing damp at the roots or moisture remaining within its saucer-shaped rosettes. Of recent years a number of members of the Society have, however, grown it successfully from seed, tempted by the challenge it presents and the fascinating symmetry of its rosettes. We have been privileged to see these specimens at the Society's Shows but, even though looking extremely healthy, they have not yet, to the best of my knowledge, been grown on to the full size characteristic of the species. Let us hope that in the not too distant future we may see this species appearing on the show bench in its mature flowering state.

## Mutatae

### S. MUTATA

I do not know whether the specific name of this saxifrage was chosen to suggest that it had arisen as a mutation, but whether or not this was the case, both the form and colour of its flowers, with their very narrow copper-clad petals, are quite strikingly different from any of the other Encrusted species.

In spite of its very unusual appearance, it is not a rare plant of localised distribution, as it occurs throughout the whole of the Alps and the Southern Carpathians being found in France, Italy, Switzerland, Austria and Roumania. Botanists distinguish two subspecies, ssp. *mutata* which answers to the type description, and ssp. *demissa* which occurs in the Carpathians and is mainly distinguished by the main axis of the inflorescence being atrophied, so that a number of lateral racemes develop to give a much lower and more spreading inflorescence only about 5 inches high. This sub-species has, however, no great horticultural distinction so will not be referred to hereafter.

The leaves are broad and strap-shaped and rather resemble some forms of *S. cotyledon*. They are, however, shiny, dark green and not lime-encrusted. The stem tends to elongate with the development of successive batches of leaves, so that the habit of the plant is rather lax. The main stem dies off after flowering and in cultivation it usually seems to have no offsets for future flowering, so that it is necessary to raise fresh stock from seed, which is freely set. In nature the production of offsets is far more frequent. The leaves are about 2 inches long and ⅓ inch wide. The flowering stem is stout and about a foot high, and bears most distinctive flowers of a peculiar coppery-yellow colour with very narrow petals.

This species is not particularly difficult to grow. It is found on limestone formations, but in cultivation lime is not essential. It is of interest as it flowers towards the end of July, later than all the other species. It flourishes best when grown in partial shade. The plant is noticeably sticky due to the presence of glandular hairs.


## Encrusted Saxifrage Hybrids

The larger number of the Encrusted Saxifrages now offered for sale are hybrids but it should be emphasised at this juncture that in many, if not most, cases they do not offer great advantage over the species. 'Tumbling Waters' and 'Southside Seedling', however, have exceptional merits. Since the species and varieties of the Euaizoonia Section cross so very easily, this list will not be comprehensive and certainly will not be as long as Dr. Clay's. Time has winnowed out the hybrids with no particular merit to commend them and I propose to deal briefly with fifteen only, all of which have some merit and are still available from commercial sources.

In each case I have given the probable parentage but this is sometimes difficult to determine with certainty, different authorities having recorded different crosses. In such cases it is, in the final resort, a matter of personal preference between the alternatives offered and it is even possible that more than two species have made their contribution.

*S.* x *burnatii* (*S. aizoon* x *S. cochlearis*)
This is a dainty little natural hybrid from the Maritime Alps which bears sprays of pure white flowers of good substance on 6 inch long, reddish stems. The leaves are narrow, silvery and lime-beaded.

*S.* 'Cecil Davies'. (Probably *S. longifolia* x either *S. aizoon* or *S. cochlearis*)
This hybrid has very compact beautiful silvery rosettes and 9 inch plumes of handsome white flowers on red stems. It makes side rosettes somewhat sparingly.

*S.* x *churchillii* (*S. hostii* x *S. aizoon*)
This natural hybrid is like a larger *S. aizoon* with conspicuously toothed and beaded leaves. The flower spikes are larger than *S. aizoon*. It is not very much grown nowadays but is still available from the trade.

*S.* x *canis-dalmatica* (*S. aizoon balcana* x *S. cotyledon*)
This hybrid has flower spikes about 18 inches high with flowers heavily spotted with purple. It is a strong growing plant with grey-green incurved rosettes.

*S.* 'Dr. Ramsey' (*S. longifolia* x *S. cochlearis*)
This is one of the very best of the Encrusted hybrids. It has particularly handsome, very silvery rosettes made up of large, broad, spoon-shaped leaves. The sprays of pure white flowers are about 9 inches in length and very decorative.

*S.* 'Esther' (*S. aizoon lutea* x *S. cochlearis*)
This is a popular and widely disseminated hybrid. With me, however, it has never proved a very striking plant as the heads of creamy-yellow flowers always seem rather heavy for the 6-inch stems which bear them and the latter flop unattractively. The fault may, however, be mine in that I have no vertical face in which to plant it. The grey-green rosettes are not particularly significant in appearance. It is free-flowering and reliable.

*S.* 'Francis Cade'
This plant is said by Mrs. Griffiths to be a hybrid of *S. longifolia*. Other authorities, including Dr. Clay, make it a hybrid of *S. cochlearis* and *S. callosa*. I incline towards the latter view. It has long, narrow, pointed leaves which are handsome in themselves and bears one-sided sprays of pure white flowers on reddish stems about 9 inches long, It is now rather difficult to find.

*S.* x *gaudinii* (*S. aizoon* x *S. cotyledon* 'Pyramidalis')
This is a natural hybrid from the Pyrenees and the Central Alps. The rosettes have the large, leathery, strap-shaped leaves of *S. cotyledon* and these widen towards the apex. The 12-inch flowering stems are widely branching almost from the base. The flowers are white, spotted with red.

*S*. 'Kathleen Pinsent' A.M. (a *S. callosa* hybrid)

When seen growing well, this is a particularly attractive hybrid on account of its unusual rose-pink flowers. It is not, however, one of the easiest of the Encrusted Saxifrages to grow and is at times a little temperamental. If happy, however, it continues to throw its 9-inch flower stems throughout the summer. The leaves are very silvery and spatulate in shape, recurving towards the tips. It was introduced by the late Clarence Elliott who showed it first in 1934 when it gained an Award of Merit.

*S*. x *macnabiana* (*S. cotyledon* x *S. callosa*)

This is a most elegant plant with a much branched inflorescence, 18 inches high, composed of very many white flowers densely spotted with red especially towards the central eye. The leaves are broad and dark green. It received a First Class Certificate in 1885 and so is quite an old hybrid. Plants sold under this name are not always the true plant with the tall 18 inch flowering stems which resulted from the original cross in the Edinburgh Botanic Garden.

*S*. x *paradoxa* (*S. hostii* x *S. crustata*)

*S*. x *paradoxa*

This is a natural hybrid found in the Eastern Alps which resembles *S. crustata*. more than *S.·hostii*. It has very narrow, silvery leaves and 6-inch spikes of rather undistinguished creamy-white flowers on dark brown stems.

*S*. x *pectinata* (*S. aizoon* x *S. crustata*)

This is a very heavily lime-encrusted natural hybrid originating in Carniola with long, narrow, dark green leaves edged with prominent serrations. The creamy flowers are borne on 6-inch flowering stems.

*S*. 'Southside Seedling' (a *S. cotyledon* cross) A.M.

This outstanding hybrid was first shown at a Royal Horticultural Society's Show on the 12 June 1951 by G. T. Sutton of Cobham. It was given an Award of Merit in May 1953 when shown by W. E. Th. Ingwersen. The citation then said: 'Above a rosette of leaves 4 inches across, the graceful flower stems, 15 inches in height, bear masses of white flowers, heavily blotched and spotted with dull

crimson, particularly towards the centre. It is claimed to be the most
heavily marked hybrid of its type. The individual flowers are ⅝
inches across and the stems and leaf stalks are covered with glandular hairs'. This variety is quite easy to grow and makes plenty of
offsets, but it likes shade for part of the day. Its second parent is
unfortunately unknown. This striking plant should be grown in
every garden.

*S.* 'Tumbling Waters' A.M. (*S. callosa* x *S. longifolia*) illustration p.49

This is another outstanding hybrid, though strangely enough it
is not in very plentiful supply in commerce, in spite of being far
from new. It was raised by Symon Lejeune in 1913, by crossing a
particularly fine form of *S. callosa lantoscana*, with the pollen of
*S. longifolia*, and was given the Award of Merit in 1920. It is very
near to *S. longifolia*, except that it makes offsets by which the plant
survives the death after flowering of the main rosette. In this respect,
it scores over *S. longifolia*. Increase by this method must, however,
be slow, or such a magnificent plant would not be so relatively
scarce and expensive. To the private grower this slowness need be
no bar and he should endeavour to keep a stock of different aged
plants.

The symmetrical silvery rosettes are made up of leaves closely
resembling *S. longifolia*, except that they are rather more strap-
shaped. The panicles of flowers are very lax and contain hundreds
of individual flowers. On this account, the plant particularly needs
to be sited so that its 30 inch long inflorescences can truly bring the
plant's name to life, as the illustration on p. 49 clearly shows.

*S.* 'Whitehill' (*S. aizoon* x *S. cochlearis* probably)

This is a neat little hybrid with dense spikes of creamy flowers on
5 to 6 inch inflorescences. This plant has dull blue-grey leaves which
are blunt ended and reddish at the base. It is a reliable, though not a
spectacular, hybrid and is widely grown and distributed.

There are two important hybrids with the Robertsoniana Section,
*S.* x *andrewsii* and *S.* x *zimmeteri*, which will be dealt with under that
Section. There is also a natural hybrid with the Kabschia Section,
*S.* x *forsteri* (*S. caesia* x *S. mutata*) as well as hybrids with *S. aizoides*.
but these are not of horticultural significance.

EDITOR'S NOTE. It was at this point that the A.G.S. *Bulletin*, from which these articles are reprinted, changed over to the metric system. We regret this apparent inconsistency.

---

## DACTYLOIDES SECTION

### (The Mossy Saxifrages)

I now come to perhaps the most difficult Section to disentangle and deal with, the Dactyloides Section, which comprises the plants known popularly under the collective term of Mossy Saxifrages. These Mossy Saxifrages are almost invariably to be found amongst the hard core of alpine plants furnishing even the most primitive of rock gardens and fully deserve the place allotted to them, for not only are they remarkably floriferous in late April and May, but their bright green mats of evergreen foliage give furnishing to the rock garden throughout the whole year. Two criticisms only can be levelled at them. The first is that they spread rather rapidly and are apt to over-run their less robust neighbours if a close watch is not kept on them. The second failing is that many of them are apt to develop brown patches in the centre of their mats if placed in full sun, particularly on the lighter soils. This habit can, however, be turned to advantage if the 'Mossies' are used in partial shade where other plants might not be so floriferous. Familiarity should not be allowed to blind us to the very real merits of these plants and, in particular, of those cultivars with rich red flowers, which are usually the least tolerant of exposure to scorching sun.

The greatest difficulty in dealing with this Section arises from the fact that almost all the 'Mossies' grown in gardens at the present time are hybrids of some complexity and uncertain parentage. The actual species which make up this interesting Section are nowadays but little grown, though a number are cultivated with some success by members of this Society in their alpine houses.

After dealing with the species I propose to list and describe very briefly the best of the hybrids now available. This list will be kept fairly short as there is little point in selecting for mention any but the current best, when so many only marginally different have been and are being produced. By the very nature of things many are of ephemeral interest as they are in due course superseded by further cultivars of real or imagined superior merits.

To Members of this Society the species inevitably have the greater appeal, both for certain intrinsic qualities associated with true species and because many present the grower with a challenge if he is to achieve success. These 'Mossy' species, with their usually white or creamy-white flowers, are not plants of breathtaking beauty and instant appeal but many are plants of character and well worth cultivating. I am sure that the fact that, amongst all the species, it is most unusual to find flowers of any colour other than white, cream or greeny-yellow has had a discouraging effect on their wider cultivation. In the case of a number, which are high alpines and inhabitants of the zones of perpetually melting snow, it may be felt that their merits are insufficient to repay the virtuosity demanded for their successful cultivation.

It is also true to say that the dividing lines between a number of the endemic Spanish species, which feature so largely in this Section, are very finely drawn and this must also have had a discouraging effect on their introduction. Nevertheless it is unnecessary to grow every species and there is a considerable number, well marked off from each other, which will supply abundant interest without recourse to those only marginally distinct.

It is a matter for regret that these species (with a few notable exceptions, such as *SS. exarata, cebennensis* and *demnatensis*) have been conspicuous by their absence from the Society's show benches and from the pages of its *Bulletin* in recent years. A considerable number have never featured at all. In the whole of the *Report* of the 1961 Rock Garden Conference there is only one mention of a plant in this Section and that is a one-line reference to *S. caespitosa* being common in Colorado! I trust, therefore, that these notes will serve to arouse interest in the large number of species which make up this Section and that a number will once again find their way into wider cultivation. It is fortunate that each year seed of a number of the species, sometimes collected from the wild, is offered in the Seed List.

Whilst the more advanced Members of the Society will, almost certainly, find their greatest interest among the species, those who are starting out along the path of alpine gardening and those whose interests remain more general, will clearly be more concerned with the 'Mossy Hybrids' which are plants of the greatest good temper, the widest possible usefulness and a colour range wider than that of the species. Their soil requirements are, in general, far from exacting, but many appreciate some shade, at any rate in the south. They like a moist soil and resent arid conditions which, whilst they will not kill them, will cause browning and burnt patches in their cushions, ruining one of their principal attractions.

When suitably placed these hybrids will rapidly make broad mats of cheerful green foliage throughout the year, which will in due course be sheeted over with flower. Where less happy, and I think this particularly applies to those with the richest red flowers, it will pay, as soon as brown areas appear in the centre of the plant, to divide and replant with rooted sections from the outside of the mat, or, if preferred, cuttings can be taken and planted out when well rooted. The propagation of these hybrids is of the simplest, whether from division, cuttings, or seed. Whilst they will not come true from seed, it can be an interesting exercise to rear a batch of seedlings in this way, when a wide variety of shades and forms will be obtained. If, however, one wishes to retain the characteristics of any named hybrid, propagation by division or cuttings is essential.

Unfortunately the cultivation of the species is, in most instances, a very different story. It must be admitted that, with certain notable exceptions, the majority are best given the protection of the frame or alpine house during the winter. It may be that a number will thrive outside if the position for them is chosen with care. They probably used to be cultivated in this way, judging from the remarks in Robinson's *English Flower Garden* but this, perhaps, is the very reason why so many have disappeared from the horticultural scene. With the advent of the alpine house, they have now been given the opportunity to stage a come-back to the extent that their merits justify.

I think it is true to say that there are, amongst knowledgeable gardeners, very considerable suspicions as to the hardiness of the Spanish species in this country and even graver suspicions of the species from North Africa and the Madeiras. I do not personally believe that, where low temperatures are concerned, these suspicions are in all cases well founded but on the other hand I am sure that most are intolerant of wet conditions at the roots throughout the winter and, under our conditions, exposure to persistent winter damp will cause the hairy-leaved species to become very bedraggled or to die completely before the advent of spring. It is not without significance that in nature they are so often found as plants of steep cliff faces, usually, in southern climates, in deep shade.

From this, I think it can be deduced that, if we can save them from being either scorched in summer or waterlogged in winter, and if we can give their foliage some protection from months of winter's damp, the difficulties involved in their cultivation will largely vanish. These are the very things which pot culture in a frame or alpine house can do. A gritty open compost will be required in the majority of cases. Whilst these fairly simple precautions will probably procure success with most species, there are some which will still remain difficult plants to cultivate, particularly the species of the Androsaceae Sub-Section.

Without going into botanical detail, it can be said that the members of this Section form a readily recognizable collection of species

73

characteristically giving rise to spreading mossy mats of neat evergreen foliage. These mats are made up of a large number of symmetrical rosettes of usually bright green leaves, entire or dissected, smooth or hairy, and often sticky to the touch as well as aromatic. In these characters they are most nearly approached by species of the Nephrophyllum Section, but the latter never give the same 'mossy' impression.

The Section is one that Engler and Irmscher divided into twelve Sub-Sections and the species which are in cultivation, or which could be brought into cultivation, are listed below under the appropriate Sub-Sections. I do not intend to go into the botanical characters of the Sub-Sections but rather to treat them as convenient groupings of species with obviously close affinities.

Professor D. A. Webb of Dublin University has, during recent years, done a vast amount of research on the saxifrages of this Section, as also of other Sections, and has published a number of revised names. In consequence, the revised nomenclature of species used in the recently published *Flora Europaea* is adopted here also.

It is worth mentioning that the foliage of the first six Sub-Sections is either entire or shortly 3 to 5 toothed at the apex, whereas that of the latter six Sub-Sections is definitely divided to a greater or lesser extent.

| | |
|---|---|
| 1. TENELLAE | *S. tenella* |
| 2. SEDOIDEAE | *S. sedoides* |
| 3. MUSCOIDEAE | *SS. muscoides* and *facchinii* |
| 4. APHYLLAE | *S. aphylla* |
| 5. ANDROSACEAE | *SS. androsacea, coarctata, depressa humilis, italica, presolanensis* and *seguieri* |
| 6. GLABELLAE | *S. glabella* |
| 7. AXILLIFLORAE | *SS. praetermissa* and *wahlenbergii* |
| 8. AQUATICAE | *S. aquatica* |
| 9. CERATOPHYLLAE | *SS. camposii, canaliculata, corbariensis, cuneata, demnatensis, geranioides, maderensis, moncayensis, pedemontana, pentadactylis, portosanctana, trifurcata* and *vayredana.* |
| 10. GEMMIFERAE | *SS. conifera, continentalis, erioblasta, globulifera, hypnoides, maweana, oranensis, reuterana, rigoi, spathulata* and *trabutiana* |
| 11. CAESPITOSAE | *SS. adenodes, boussingaultii, caespitosa, hartii, lactea, magellanica, pavonii, rosacea* and *sileniflora* |

## 12. EXARATO-MOSCHATAE SS. *cebennensis, exarata, hariotii, moschata, nervosa, nevadensis,* and *pubescens.*

### Tenellae

#### S. TENELLA

This species is an inhabitant of shady rocks and screes in the South-Eastern Alps of Austria, Italy and Yugoslavia. It makes an attractive plant with its neat mats of fine emerald-green foliage and slender 10 cm. high inflorescences bearing up to eight creamy-white flowers. Cultivated in this country for many years it still features in at least one nurseryman's list. Though not showy, it is nevertheless well worth growing. It is not one of the more difficult species and in a suitable moist shady corner it will grow well in the open garden.

### Sedoideae

#### S. SEDOIDES

This small mat-forming species is a high alpine of the limestone ranges in the Apennines, Eastern Alps and the mountains of Yugoslavia and Albania. It is a plant of the moist screes and forms a rather loose, low mat of rosettes of entire leaves. The flowers are greenish-yellow and are borne two or three on a short flowering stem. It is not a particularly easy plant to please in cultivation at lower altitudes and, on account of its rather dowdy appearance, it is seldom grown. It is divided into two subspecies—ssp. *sedoides* and ssp. *prenja,* the latter being distinguished by some of its leaves being 3-toothed.

### Muscoideae

#### S. MUSCOIDES

This is another very dwarf, compact, mat-forming species, which is a little easier to grow in the garden than the last species. It occurs above 2,200 metres throughout the Alps and is found in Italy, Switzerland, France and Austria. The leaves are quite entire and turn a silvery colour when they die and dry off. The white or yellow flowers are borne singly, or up to three on a short flowering stem. Many plants cultivated as varieties of *S. muscoides* are almost certainly varieties of *S. moschata,* as there has been much confusion between the two species in horticultural circles.

#### S. FACCHINII

This species closely resembles the last, but has even smaller, tighter cushions. The pale yellow flowers, sometimes tinged with purple, are borne two or three to a very short flowering stem. It is a local plant of the Southern Dolomites, found above 2,200 metres, and so occurring only in Italy. It is seldom cultivated and, though of neat habit, its flowers are even more insignificant than the last.

**Aphyllae**—see note on page 93.

# Androsaceae

### S. ANDROSACEA (illustration p. 78)

This is a very widely-spread tufted high alpine saxifrage found on all the mountain ranges of Europe from the Pyrenees to the Carpathians. It tends to occur in acid conditions on non-calcareous formations. The leaves, which are 1 to 2 cm. long and up to ½ cm. wide, are entire or sometimes very shortly 3-toothed. It is a definite moisture lover, and can only be made happy with moraine conditions. 1 to 3 creamy-white flowers are borne on 4 to 8 cm. flowering stems and the inflorescence is quite ornamental in a mild way. The species has a certain charm and is worthy of cultivation if only its rather specialized moisture requirements can be met. Seed is often offered in the Seed List.

### S. SEGUIERI

This species is confined to the higher zones of the Central and Eastern Alps and resembles a smaller *S. androsacea*. The leaves, however, are always entire, the flowers dull yellow, and the flower stems no more than 5 cms. in height. Like *S. androsacea* it has a certain modest charm, but is almost certainly equally difficult to accommodate happily.

### S. ITALICA

This species, until recently called *S. tridens*, is only found in the Abruzzi Mountains in the Apennines, where I have seen it making neat mats of dwarf rosettes pressed closely to the ground. It resembles a dwarf *S. androsacea*, but the leaves are 3-cleft and it is always very compact with few-flowered inflorescences about 3 cms. high. The flowers are milky white and relatively large for the size of the plant. I have not cultivated this plant nor seen it in cultivation.

### S. DEPRESSA

This is another species of the *S. androsacea* persuasion which occurs only in the Southern Italian Alps and bears 3-lobed leaves. Its flowering stems are slightly longer than the last species, i.e. 5 to 10 cms. and it is said to inhabit shady ledges and damp screes on porphyritic volcanic rock between 2,000 and 2,850 metres. 3 to 8 rather small and not very attractive flowers are borne on each inflorescence. This is not a species that has any great horticultural merit, though collected seed has been offered in recent years in the Seed List.

### S. PRESOLANENSIS

This species is a taller, more slender version of *S. androsacea* with flowering stems up to 12 cms. in height, bearing 2 to 4 small greenish-yellow flowers. It occurs only in one locality in the Bergamasche Alps in Northern Italy. This is another species with small claim to be introduced into cultivation.

*Saxifraga geranioides* (p. 85)     *Photo: M. G. Hodgman*

*Saxifraga androsacea* (p. 76)

*Photos: M. G. Hodgman*

*Saxifraga pedemontana* (p. 82)

*Saxifraga demnatensis* (p. 85)        *Photo: Roy Elliott, A.R.P.S.*

*Saxifraga oppositifolia alba* (p. 100)

*Photos: R. Elliott*

*Saxifraga cuscutiformis* (p. 112)

### S. HUMILIS
This species occurs in Alpine and Sub-Alpine Yunnan and has not been brought into cultivation.

### S. COARCTATA
This species occurs in the Alpine and Sub-Alpine Himalayas and has not been brought into cultivation.

## Glabellae

### S. GLABELLA
This is a species of the Eastern Alps and the Abruzzi Mountains, with completely glabrous and entire, narrowly spatulate leaves. The flowering stems are up to 10 cms. in height and bear several small white flowers of no great distinction. It is found on limestone screes and has probably no great claims to garden cultivation.

## Axilliflorae

### S. PRAETERMISSA

This is the species which, until recently, was called *S. ajugifolia*. It occurs only in the Pyrenees in the Cordillera Cantabrica above 1,500 metres and is often found on the stony margins of lakes. The plant forms weakly spreading, leafy shoots, characteristically rather deeply divided into 3 or 5 parts to give the foliage a cut appearance. The flowering stems arise from the axils of the leaves on the lower part of the leafy stems and reach a height of up to 12 cms. bearing 1 to 3 medium-sized white flowers. The species is in cultivation and I am presently growing it from seed offered in the Seed List. It promises to be quite an ornamental plant worthy of cultivation. It obviously needs moist, but not water-logged conditions.

### S. WAHLENBERGII

This species, formerly known as *S. perdurans*, comes from damp grassy slopes of the Western Carpathian Mountains in Poland and Czechoslovakia. Like the previous species, its leafy stems are decumbent and distinctly cut into 3 to 5 segments with the flowering stems arising from the axils of the leaves on the lower part of the leafy stems. The flowering stems are shorter than the last, and again bear up to 3 medium-sized white flowers. This species appears to be in cultivation for seed of it, not marked as having been collected, has been offered in the Seed List in recent years, but I have not seen it grown. It would appear to be worth trying.

## Aquaticae

### S. AQUATICA

This species is found on the margins of streams in the Eastern and Central Pyrenees, growing from 1,500 metres upwards. It is a strong-growing plant with stout, erect flowering stems up to 60 cms. in height and grows in large masses where conditions are to its liking. The shiny leaves are large, leathery and deeply divided, giving the impression of Buttercup leaves. The flowering stem bears at its top a narrow compact panicle of numerous starry white flowers, which are relatively large and handsome. This is probably not a very difficult plant to grow where it can be given a suitable streamside position, with its roots reaching down to the water. Collected seed is offered from time to time in the Seed List and it is worth trying where one can suit its requirements. It is quite a handsome summer-flowering species.

## Ceratophyllae

### S. PEDEMONTANA (illustration p. 78)

This species is characteristic of this Section. It is widespread and very variable, occurring from the Pyrenees to the Carpathians and the mountains of the Balkans. It breaks down into four subspecies:

ssp. *pedemontana*
This occurs in the South Western and Central Alps. The leaves are thick and fleshy and the segments are all forwardly directed with a broad petiole.

ssp. *cymosa*
This occurs in the Eastern Carpathians and the mountains of the Balkans. The leaves are thin and soft and again the segments are all forwardly directed with a broad petiole. It is altogether a smaller and more compact plant with close clusters of flowers on 8 cm. flowering stems.

ssp. *cervicornis*    This occurs in Corsica and Sardinia and is particularly distinguished by the rosette leaves being incurved in the bud, having only a short petiole and the segments are spreading.

ssp. *prostii*

This occurs in the Cevennes and the Pyrenees. It has large, fleshy, wedge-shaped, glandular leaves not incurved in the bud and the lateral segments are often spreading with a narrow petiole.

All the sub-species of this plant seem to prefer non-calcareous rock formations. They are not difficult plants to grow, given a certain amount of care in placing them in a moist, shady position where they will not be waterlogged in the winter. In the alpine house, they are quite easy and rather showy with handsome, fairly large white flowers.

### S. PENTADACTYLIS

This species hails from the Eastern Pyrenees and the mountains of Central and Northern Central Spain. It is in the same group of smooth-leaved saxifrages as *SS. corbariensis, trifurcata, canaliculata* and *camposii* and is distinguished among them by the combination of the 3 to 5 blunt leaf segments and the only moderately lengthy petiole. The leaves are sticky and aromatic. In nature it is found on the non-calcareous rocks. The much-branched flowering stems bear 3 to 30 small, starry, white flowers tinged with cream. This is a moderately interesting plant for alpine house culture, but unlikely to survive for long in the rock garden. Seed is frequently offered in the Seed List, so there is no difficulty in giving it a trial.

Var. *willkommiana* is a form occurring in Central Spain which makes a rather more robust and striking plant.

### S. CORBARIENSIS

This species comes both from the mountains of Corbières at the eastern end of the Pyrenees and from Eastern Spain, between Teruel and Jaen. There are differences between the populations from the two areas, those from the former ranking as ssp. *corbariensis* and those from the latter as ssp. *valentina*. The leaves are boldly divided, into 7 to 11 segments in ssp. *corbariensis* and reputedly 5 segments (the central lobe being entire) in ssp. *valentina*. In both, the leaves are characterized by their very long petioles which give the foliage a distinct and easily recognizable appearance. The leaves are completely glabrous and bright green. In ssp. *corbariensis* the flowering stems are said to be up to 25 cms. high and bear large, handsome, pure white flowers. When I have found it in the Corbières, the flower stems have only been 10—12 cms. high and all the leaves had 5 to 7 segments. It has proved an amenable plant for the frame and alpine house, but I have not tried it in the rock garden.

### S. TRIFURCATA

This species, which was at one time known as *S. ceratophylla*, grows only in Northern Spain from Asturias to Navarra. Unlike some of the species in this Sub-Section, it is not difficult to grow and its dark green leathery leaves will withstand the sun well. It is not demanding with regard to positioning. The leaves are sticky and much divided, their mode of division giving them a superficial resemblance to a Stag's Horn. The panicles of large and handsome white flowers are carried in a spreading manner with the flowers all facing upwards. The species is exceedingly floriferous and is a plant well worth growing in the garden.

### S. CANALICULATA

This species comes from the Cordillera Cantabrica, which are limestone mountains in Northern Spain. It forms broad cushions of narrowly divided, very sticky leaves giving a superficial impression of lichen-like foliage. The segments of the leaves and the petiole are all deeply grooved on their upper surface, which gives the plant its specific name. It carries abundant trusses of large white flowers on stems up to 15 cms. in height and gives a handsome effect. This is a species well worth growing, but probably only as an alpine house plant. It is not commonly seen in cultivation.

### S. CAMPOSII

This species hails exclusively from the limestone mountains of South Eastern Spain, from the vicinity of Ronda to the Sierra de Alcarez. It forms neat compact cushions of slightly sticky, much-divided leaves with flowering stems up to 15 cms. in height bearing dense clusters of handsome white flowers. This is a rare species and I doubt whether it is now in cultivation in this country, except in the collection at Kew. It should be grown again, but it would probably require alpine house cultivation.

### S. VAYREDANA

This species occurs only in the Mountains of Montseny, north-east of Barcelona in Spain. It forms a distinctly rounded cushion which, when in flower, is almost hidden by the wealth of bloom. The flower stems are very slender and only 6 to 10 cms. high, being reddish in the upper part. Each stem carries 6 to 7 flowers which individually are small. The foliage has a characteristically mossy appearance, individual leaves having a long petiole and are 3-cleft at the apex. The petals are prominently veined and are broad but not overlapping. The flowers appear to be larger in certain individuals than others. This species makes a valuable alpine house plant at Kew where it makes a regular appearance in mid-June. It is well worth cultivating for its floriferousness and its season of flower.

### S. MONCAYENSIS

This plant from the Sierra del Moncayo in North Eastern Spain was only established as a species in 1963 by Professor D. A. Webb who wrote, "It is intermediate in most features between *S. pentadactylis* and *S. vayredana* differing from *S. pentadactylis* in the larger cushions of soft, lighter-coloured foliage, the indumentum of numerous, very short, glandular hairs and the slightly larger petals; from *S. vayredana* in the linear, strongly sulcate, leaf segments up to 10 mm. long and the much fainter scent". [*Flora Europaea I.*]

This plant can be seen at Kew but I do not think that it is in general cultivation. I doubt whether it is as potentially useful horticulturally as *S. vayredana*.

S. CUNEATA

This species inhabits the Western Pyrenees and the calcareous mountains of Northern Spain. Its habit is similar to *S. pentadactylis*, but it is easily distinguished from the latter by its wedge-shaped leaves which are thick, glandular and sticky. The pure white blossoms are moderately large and beautiful and are borne on flowering stems up to 15 cms. high. I have not seen this plant in cultivation but, if attempted, it should be grown in the alpine house.

S. GERANIOIDES (illustration p. 77)

This species from the Pyrenees and North-Eastern Spain has flowering stems up to 20 cms. high, bearing a compact head of 6 to 9 pure white campanulate flowers. The long-petioled leaves are very much divided and have a superficial resemblance to those of certain *Geranium* species. With some shade and a moist, but not waterlogged soil, this is a pretty species, well worth growing. Seed has been offered in the Seed List in recent years.

S. DEMNATENSIS (illustration p. 79)

This species comes from the Atlas Mountains where it was discovered by E. K. Balls. It was re-introduced into cultivation by Archibald from his expedition to North Africa in about 1963 where he collected it from below the western summit of Djebel Toubkal. *S. demnatensis* gained a Preliminary Commendation when shown by Roy Elliott in April 1965, and is described and photographed in the December 1965 issue of the *Bulletin* (Vol. 33 pp. 352, 357). It forms a mat of pale green, sticky, aromatic leaves covered by glandular hairs. Whilst on the plant shown the inflorescences were only 5 to 8 cms. high and rather compact, they appear to rise to 12 cms. or perhaps even more, when growing in shade in the wild, and the form of the inflorescence is then looser. The latter carries at least 10 large pure white flowers, Archibald describing the inflorescence as 'billowing'. With me this species is just reaching flowering stage and to date it has grown well both in a frame and in the alpine house.

S. PORTOSANCTANA

This species comes from the little island of Porto Santa in the Madeiras, where it grows on the north-facing slopes at a height of 500 metres. Its thick fleshy leaves are light green in colour with only a medium length petiole. Apart from the length of the petiole, the 3-cleft leaves are reminiscent of *S. corbariensis*. The inflorescences of 2 to 3 flowers are carried on stems 10 to 12 cms. in height, these stems being markedly zig-zag. The flowers are large, pure white and prominently veined. This is a handsome plant but, coming from the latitude that it does, it should definitely be given the protection of the alpine house, where it flowers in June.

S. MADERENSIS

This species, which is found at an altitude of about 1,000 metres in Madeira, is somewhat more tender than the last. It is rarely, if ever, seen in this country. Its general appearance is uncharacteristic of this Sub-Section, for the leaves are kidney-shaped and cut into 5 lobes, each of which is notched 3 to 5 times. It bears large spreading panicles of 5 to 7 handsome white flowers on 10 to 15 cm. high stems. It is worth attempting if the alpine house is frost proof.

## Gemmiferae

S. HYPNOIDES

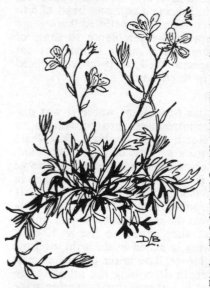

This is a widely-spread and variable species, well-known in this country, occurring as it does as a native plant, commonly known as the 'Dovedale Moss'. It ranges from the Sub-Arctic regions through North Western Europe, extending as far to the south east as the Vosges. It is a loosely knit, prostrate plant which sends out straggling leafy shoots bearing entire linear leaves and ending in a rosette of 3 to 7 lobed leaves. Summer-dormant buds usually occur in the lower leaf axils, a characteristic of this Gemmiferae Sub-Section. The flowering stems are normally about 10-15 cms. in length and bear 3 to 7 nodding buds which open out into fairly large, pure white, starry flowers. This is a very good evergreen rock garden plant which provides clothing throughout the year, though its place has largely been taken over by the various 'Mossy Hybrids' whose flowers are fuller and available in a wider range of colours.

'Kingii' is a very dwarf cultivar of S. hypnoides which forms a delightfully compact moss-like carpet, the foliage of which takes on distinctly red tints in the autumn.

'Whitlavei' is another cultivar, similar to 'Kingii' but is not so dwarf and does not redden in the autumn. It is, however, considerably more floriferous and bears more flowers to each inflorescence.

86

## S. CONTINENTALIS

This used to be considered a sub-species of *S. hypnoides*, but has now been given specific rank. It has a more southerly distribution, being found in Southern France, Northern Spain and Northern Portugal. It differs mainly in the stalked summer-buds, which are always present in many of the leaf axils, and which are covered by completely scarious silvery leaf-scales. It has no particular horticultural advantage over *S. hypnoides* and is probably more difficult to grow.

## S. CONIFERA

This species is confined to the Cordillera Cantabrica in Northern Spain. It is a very small compact 'Mossy' bearing numerous dormant buds which dry up in summer. The leaves are linear, undivided and pale silvery-green, and the 3 to 5 small white flowers are borne on 6 to 10 cm. flowering stems. In the *Bulletin* (Vol. 2 p. 212) Dr. P. L. Giuseppi described finding it on damp limestone cliffs on the Picos de Europa.

It is probably not an easy plant to grow, though I note it was offered in the 1967 and 1968 Seed Lists. I have not yet tried it, but intend to do so, for, though it is obviously not a spectacular plant, it would seem to be a plant of some character.

## S. GLOBULIFERA

This species is found on the limestone rocks and cliffs of Southern Spain. Like other members of this Sub-Section it has a spreading mat-like habit, with flowering stems about 7 to 12 cms. long and numerous summer-dormant axillary buds. The leaves are semi-circular in outline, but deeply 3 to 7 lobed: characteristically, the lower leaves have long leaf stalks. The inflorescence consists of 3 to 7 rather small white flowers. Plants labelled as *S. globulifera* var. *erioblasta*, *spathulata* or *oranensis* are sometimes seen in cultivation but all these three have now been raised to separate specific rank and are described below.

*S. globulifera* should be grown as an alpine house plant, if attempted, but it would not appear to have particularly attractive qualities.

## S. ORANENSIS

This species, as mentioned above, has close affinities with the Spanish *S. globulifera*, but is actually found in the Djebel Santo Mountains in the neighbourhood of Oran in Algeria. The main difference is that it has much larger and more decorative flowers and its leaves turn red in summer after the flowering period. The fleshy wedge-shaped leaves are only faintly notched at the apex. In the December 1948 *Bulletin* (Vol. 14, p. 317), F. W. Peacock describes growing it and was enthusiastic about its merit and interest as an alpine house plant. Dwight Ripley in the 1937 *Bulletin* (Vol. 5, p. 247), also has a good word to say for it, rating it as an easier plant to grow than the Spanish *S. erioblasta* or *S. conifera*. It must be kept on the dry side during the winter.

## S. SPATHULATA

This is another Northern African species occurring in Algeria. It can be recognized by its narrow undivided spatulate leaves, its very lax habit and its relative lack of hairiness. The flowers, which are borne in sprays on stems up to 20 cms. in length, are neither very large nor very decorative. This is not a very attractive species and is not often seen in cultivation, though I notice that seed was offered in the 1968 Seed List. It obviously needs careful alpine house cultivation and even then might prove a short-lived plant.

## S. ERIOBLASTA

This Spanish species, coming from the Sierra Nevada Mountains in Southern Spain, protects itself during the hottest period of the year by forming tight summer-dormant buds. In Vol. 2 of the *Bulletin*, Dr. Giuseppi contributes some notes on it, as he found it on his September 1933 Spanish expedition. He says, 'The plant dries up in the summer drought into small grey buds which, from their resemblance, are called 'Virgin's Pearls' by the Spaniards. If these little buds are crushed, a tiny green spark of life is found in the centre of each. The flower is small and white and the minute leaves are slightly lobed'. In the December 1948 *Bulletin* (Vol. 16, p. 317), F. W. Peacock describes having grown it with interest and says that careful culture in the alpine house is essential. It is an extremely compact little plant, with 3 to 5 flowers on the 5 cm. flowering stems.

## S. MAWEANA

This species was discovered and brought to England from near Tetuan in Spanish Morocco in 1869 by Mr. George Maw. It was collected by him again in company with Hooker and Ball in 1871. An enthusiastic description of its horticultural potential appeared in the *Gardeners' Chronicle* in 1871 which described it as 'a member of the hypnoides group with flowers as large as *S. granulata*' and 'nearest to the Algerian *S. oranensis* which, however, has more numerous small flowers and more rhomboidal leaves'.

The leaves are fleshy and 3-lobed, each lobe being itself 3-lobed, and the petioles are 2 to 3 times as long as the leaf blade. The new shoots are purple in the lower half and axillary summer-dormant buds are produced profusely. 4 to 9 large white flowers are carried in a lax corymb on a flowering stem 10 to 15 cms. in height.

There is no doubt that this is a choice and attractive plant, but, as far as I can ascertain, it appears now to have disappeared from cultivation in this country. It is well worth re-introduction though it is probable that, coming from as far south as it does, it will require alpine house or frame treatment.

## S. TRABUTIANA

This is a rare species from the Djebel Magris Mountains in Algeria. It is botanically very close to *S. erioblasta*, but is not such a compact plant. It carries an inflorescence of 3 to 5 small white flowers on a 5 to 6 cm. long stem. This species does not appear to be in cultivation in this country and hardly seems worthy of the alpine house culture which it would assuredly need.

### S. REUTERANA

This species comes from Granada and the mountains of Southern Spain. It is very like *S. globulifera*, but has shorter flowering stems, only 4 to 6 cms. in length, and the flowers are fewer and larger, only about 1 to 2 to the inflorescence. In cultivation, however, it is not unusual for it to produce 4 to 5 relatively large pure white flowers. This plant is grown in the alpine house at Kew, where it is quite ornamental but it is doubtful whether it is in cultivation elsewhere in this country.

### S. RIGOI

This is yet another species from Granada. Dr. Giuseppi, in the *Bulletin* (Vol. 2, p. 212) describes finding it growing on the bare limestone cliffs of the Sierra Corzola where, he says, the plant dries up in the drought of the summer, becomes red, and so forms huge red clumps which are a great contrast to the white of the cliffs.

The leaves are very divided and the flowers are large and white, borne 2 to 3 on a 6 cm. flowering stem.

The species is rather like *S. globulifera*, but for horticultural purposes an improvement on it because of the larger flowers and the compact and free-flowering habit. It seems to have gone out of cultivation in this country, but is well worth re-introduction as a handsome alpine house plant.

### Caespitosae

### S. CAESPITOSA

Although the name is often used horticulturally, the true species is but rarely seen in cultivation in this country, most of the forms referred to it being hybrids with other species. It is a plant of the

Arctic and extreme Northern Regions being found, rarely, near the summits of one or two mountains in Wales and Scotland. It occurs plentifully in the Faroes, Iceland, Greenland and North America. Seed collected in North America is frequently offered in the Seed List. It is somewhat like *S. hypnoides*, but is a compact tufted plant without any loose runners. The foliage is usually 3-lobed and not unlike that of many of our common 'Mossy Hybrids' but it is certainly not as good-tempered as the latter in cultivation in this

*S. caespitosa*

country. As the flowers, which are borne 1 to 3 on a 2.5 to 8 cm. stem, are of a rather impure white colour it is not a very attractive species, except possibly when seen in masses in its native haunts. Its main importance is as one of the original parents of our very mixed-up 'Mossy Hybrids'.

There is also a naturally occurring hairy variety called var. *hirta* which is quite distinctive in appearance.

## S. ROSACEA

The next most widely spread of the saxifrages in this Section is *S. rosacea* formerly known as *S. decipiens*. This is a plant of North-Western and Central Europe ranging from Poland to Iceland through Germany, France and Great Britain. It is an extremely variable species, some forms making very compact cushions, whilst others spread into wide carpets.

The leaves are usually rather bluntly 5-lobed and the 3 to 5 pure white flowers are borne on a flowering stem varying in length up to 25 cms.

It is certainly one of the most important parents of the 'Mossy Hybrids' and its influence is clearly seen in them.

The only sub-species satisfactorily and constantly distinct from the type is ssp. *sponhemica* which has sharply pointed tips to the lobes of the leaf segments.

## S. HARTII

This species is intermediate in many ways between *S. caespitosa* and *S. rosacea* and is found in only one isolated locality, the maritime cliffs of Arranmore Island in North Western Ireland. It is quite distinct and rather attractive in appearance. The leaves are very hairy and sticky to the touch. The fairly large, pure white flowers are borne in a very compact inflorescence on a 4 to 5 cms. high, particularly sturdy flowering stem. The flowers are given an unusual and distinctive appearance by the upper half of the sepals being red.

| | |
|---|---|
| S. ADENODES | South America, Chilean Andes. |
| S. BOUSSINGAULTII | South America, Ecuador. |
| S. LACTEA | Asia, Sub-Arctic Asia. |
| S. MAGELLANICA | South America, Magellan Straights and Peru. |
| S. PAVONII | South America, Argentine and Chile. |
| S. SILENIFLORA | North America, Behring's Land, Sub-arctic America, Alaska. |

The above species are all described by Engler and Irmscher as members of this Section with a greater or lesser resemblance to *S. caespitosa*. I have not seen them, so I will pass over them with only the above reference. It seems doubtful whether they are very likely to have much significance for gardeners in this country, though they might display distinctive qualities in other climates. It would, however, be interesting to try them if seed became available.

**Exarato-moschatae**

S. EXARATA (illustration p. 47)

This is a very widely spread and variable species of the high mountains of Southern Europe, where it occurs in the Pyrenees, the Jura, the Alps, the Apennines and the Balkan Mountains. It resembles *S. moschata* in many ways.

The leaves are very variable in size and shape, but are most usually 3-lobed and, in contrast to *S. moschata*, have a distinct nerve or channel running down the centre of each strap-shaped segment. The flower stems are 3 to 10 cms. tall, bearing 3 to 8 creamy-white flowers.

This species is often seen at the Shows, but the position is somewhat confused by the fact that some, at any rate, of the plants exhibited under this name are *S. pubescens* ssp. *iratiana* which will be described subsequently. *S. exarata* is not an easy species to keep going for long and it must certainly have alpine house culture. It makes, however, a neat and attractive plant and is worth a little trouble.

S. MOSCHATA

This mat-forming species, which is rather similar to the last, occurs over the same wide range of Southern and Central Europe from the Pyrenees to the Carpathians. It is a very variable species and, to confuse matters still more, it hybridizes freely with *S. exarata* and *S. pubescens*. The light green leaves are usually 3-cleft like *S. exarata* but, unlike the latter, the lobes are without prominent veining or channelling. The flowering stems are from 1 to 10 cms. long and bear 1 to 7 flowers, which are usually dull yellow or cream in colour, though they may be tinged with deep red or even nearly white. It is not a difficult species to grow in the open garden.

Very many different forms have been given distinctive names such as vars. *pygmaea*, *compacta*, *lineata*, *atropurpurea* and *laxa* but these, though occurring naturally in different areas, are mostly difficult to find in cultivation, having now been supplanted by various more showy hybrids containing *S. moschata* blood. This species has certainly contributed to the popular red-flowered hybrids. One cultivar which is said to have originated from this species in cultivation is *S. moschata* 'Cloth of Gold', which has striking golden foliage throughout the year. It is readily available in commerce, though it is a plant which needs placing away from direct sun if it is to flourish. It is not at all robust and to be grown outside it needs a special position such as a shady trough.

S. PUBESCENS

This species, which is very nearly akin to *S. exarata* and *S. moschata*, occurs only in the Pyrenees. The foliage is rather dark green and usually 5-cleft though sometimes 3-cleft, the leaves being densely covered with long glandular hairs, a feature by which the species can be recognized. There are two distinct geographical sub-species, ssp. *pubescens* and ssp. *iratiana*.

### ssp. *pubescens*

This occurs mainly in the Eastern Pyrenees and has leaves 10 to 20 mms. long with the segments four times as long as wide. The flowering stems are usually 3 to 10 cms. tall and the flowers are pure white.

### ssp. *iratiana*

This occurs mainly in the Central Pyrenees and has leaves only 4 to 10 mms. long, with segments twice as long as wide. The flowering stems are usually 2 to 6 cms. tall and the white flowers are almost always veined with red. The red anthers also form a very noticeable feature of the flower. Ssp. *iratiana* is usually a more compact plant than ssp. *pubescens*.

Both sub-species are currently grown by members and have been shown on a number of occasions in recent years. Ssp. *iratiana* is, in my opinion, the more attractive plant. Neither is too difficult if grown in the alpine house in an open gritty compost, but they are definitely not plants for the garden in this country. Seed of both has been offered in the Seed List recently.

### S. CEBENNENSIS

This species grows only in the Cevennes in Southern France, where it is fairly common. The bright green leaves are almost always 3-cleft and the rosettes form a close dome-like cushion of rather characteristic appearance. The sticky leaves are covered by distinctly glandular hairs. The flowering stems are about 5 cms. long and carry about 3 full-petalled pure white flowers which are very attractive. I have not found this a particularly difficult plant, either in the alpine house or frame, but have not ventured to grow it in the open ground. It has been fairly widely grown by members of the Society in recent years and has regularly featured in the Seed List.

### S. NERVOSA

This species, which comes close to *S. exarata*, is found on basic rocks in the Central Pyrenees. It forms small loose cushions of dark green hairy leaves, the glandular hairs covering also the pedicels and calyces. The scented leaves are 3 to 5-cleft. The flowering stems are 4 to 10 cms. long and bear 3 to 12 white rounded flowers. I have neither seen nor grown this species, so can say nothing about its horticultural value.

### S. NEVADENSIS

This species from the Sierra Nevada in Southern Spain is very close to *S. pubescens*. It is, however, distinct from the two sub-species of the latter in its usually 3-cleft non-channelled leaves. Also the petals are usually narrower and sometimes red-veined. Again I know nothing of the horticultural value, if any, of this species.

## S. HARIOTII

This is a lime-loving species from the Western Pyrenees, whose shiny, nearly glabrous 3-cleft leaves are markedly different from the other species in this Sub-Section. The leaves are channelled. The flowering stems are 2 to 6 cms. high, bearing 1 to 4 flowers, whose petals are dull yellow or creamy in colour. I am in the process of growing this species from seed offered in the Seed List and to date it does not seem temperamental, though it has not yet reached flowering stage. It is, therefore, too early for me to pronounce on its horticultural merits. The colour of the flowers does not sound very interesting.

CORRIGENDUM. Owing to an error, the Aphyllae Section was omitted from page 75. It contains only one species, *Saxifraga aphylla* from the European Alps, which bears solitary yellow flowers and is of little or no horticultural value.

------

# MOSSY SAXIFRAGE HYBRIDS

As these hybrids are so numerous and confusing I propose, instead of treating them in one alphabetical sequence, to separate them into a number of groups according to their flower colour and habit and then to deal very briefly with the qualities of the individual hybrids or cultivars in each group.

### Tall, Carmine, Crimson or Scarlet

'DUBARRY' This fine hybrid carries 7 to 9 large crimson flowers on 20 to 25 cm. stems. It makes a broad mat and flowers relatively late.
'FOUR WINDS' This is a hybrid of outstanding merit. Its brilliant, deep crimson flowers are 2 to 3 cms. in diameter and are borne in clusters of four to five on a 20 cm. stem.
'POMPADOUR' This is one of the older hybrids, but is still worthy of a place in any collection. It has very large dusky-crimson flowers on 20—25 cm. stems. It is a strong spreading plant, whose habit is unfortunately not as compact as some.

### Pink

'DIANA' This is an early flowering hybrid with light pink flowers on 15 cm. stems. It has a moderately spreading habit.
'GAIETY' This is another early flowering hybrid with flowers of deep pink. It is an attractive plant with a neat habit. The flower stems are about 10 cms. high.
'WINSTON S. CHURCHILL' This hybrid bears large flowers of an attractive shade of clear pink on 10 to 15 cm. stems.

## Yellow

'FLOWERS OF SULPHUR' This hybrid is unique in possessing pale lemon-yellow flowers of moderate size. The habit is spreading and the stems are approximately 10 cms. high.

## Medium height, Carmine, Crimson or Scarlet

'BALLAWLEY GUARDSMAN' This recently introduced hybrid has flowers of a velvety crimson-scarlet of a rather subtle shade. The flower stems are about 15 cms. high.

'CARNIVAL' The flowers of this hybrid are crimson-rose and are borne on 12 to 15 cm. stems. It makes a very good compact plant.

'SANGUINEA SUPERBA' This old hybrid, with its deep crimson non-fading flowers still deservedly retains a place in most lists. Its flowers, which are freely produced and last well, are carried on 12 to 15 cm. stems.

'SPRITE' This neat hybrid has crimson-rose flowers borne on 10 cm. stems and is an attractive plant.

'TRIUMPH' This is a hybrid of outstanding merit with large non-fading, blood-red flowers borne on 15 cm. stems. The foliage makes a good compact mat.

## Low, Carmine, Crimson or Scarlet

'ELF' This is a delightful miniature 'Mossy' with carmine flowers borne on 5 to 8 cm. stems. It is a comparatively late-flowerer.

'GNOME' This is another late-flowering dwarf hybrid with crimson flowers on 5 to 8 cm. stems. It is very neat and of equal merit with the foregoing.

'PETER PAN' This is yet another very compact hybrid with crimson flowers on 5 to 8 cm. stems.

'MRS. PIPER' This hybrid has red flowers on 5 to 8 cm. stems.

'PIXIE' This is an exceptionally neat and diminutive plant with rose-red flowers on 5 cm. stems.

## White

'FAIRY' This hybrid is said to be a form of *S. moschata* and has certainly diverged less from that species than many of the hybrids. It is very compact, late flowering and low, having flowering stems about 7 to 8 cms. high.

'JAMES BREMNER' This hybrid, dating back to about 1930, is of a rather different type from more recent productions. It is tall, 20 to 25 cms. in height, has huge white flowers, $2\frac{1}{2}$ cms. in diameter, and is rather diffuse in habit. Some authorities suggest that *S. granulata* features in its parentage. It is a strong grower needing plenty of space.

'KINGII' This has already been referred to as a cultivar of *S. hypnoides*. Whether a hybrid or not, it is a plant of exceptionally prostrate habit with very small rosettes. It produces its not very distinguished white flowers somewhat sparingly.

'PEARLY KING' This is a neat and free-flowering hybrid with clear white flowers on 8 cm. stems. It is an attractive plant.

'WHITE PIXIE' This is a very neat little plant resembling 'Pixie' in all except the colour of the flowers. It is diminutive and has flowering stems only 3 to 5 cms. in height.

'WHITLAVEI COMPACTA' or 'WHITLAVEI' This has already been referred to as a cultivar of *S. hypnoides*. It is a slightly less dwarf and more floriferous version of cv 'Kingii' whose foliage does not turn red in autumn. It is a useful carpeting plant.

---

## PORPHYRION SECTION

THE THREE SPECIES BELONGING TO THIS SECTION, have a very characteristic appearance. They differ from all others, except certain unintroduced Himalayan Kabschia species and those of the Tetrameridium Section, in having opposite leaves. Of the latter Section only one species has ever been introduced to cultivation, and that is devoid of petals. The members of the Porphyrion Section are mat-forming plants with small leaves and, once seen, they are unlikely to be confused with any other species.

Whilst not challenging the horticultural importance of the extensive Kabschia, Euaizoonia and Dactyloides Sections, the Porphyrion species have a definite claim to horticultural fame, stemming mainly from the many valuable forms of that old favourite, *S. oppositifolia*. Of the other members, *S. retusa* and *S. biflora*, the former is a most beautiful and interesting alpine house plant, but the latter has only an extremely limited horticultural value.

I have said that the Section contains only three normally accepted species, but this number is often expanded to four, five, six or even seven species, according to the author's interpretation of the limits of the species *S. oppositifolia*. I have, however, followed the decisions of Engler and Irmscher in their monograph and of Professor D. A. Webb in *Flora Europaea*, Vol. 1.

The species in this Section range in difficulty of culture from some forms of *S. oppositifolia*, which are exceptionally easy plants to grow in the rock garden, through *S. retusa*, which demands rather exact conditions of soil, drainage and water supply, to *S. biflora*, which is undoubtedly difficult to grow with any lasting degree of success. Generally speaking, they like a porous gritty soil, a reliable supply of moisture, good drainage and a cool but open position where the soil above their roots will not be baked hard by the sun.

The propagation of these species is easy, as all strike readily from cuttings taken after flowering and inserted in a sharp rooting medium. As they are mat-forming plants, usually rooting as they go, it is often possible to detach pieces possessing some roots and to plant them as 'Irishman's cuttings'.

## S. OPPOSITIFOLIA L.

This species occurs over a vast area, including the sub-arctic regions of North America, northern Europe and northern Asia. In North America, its range extends southwards through the Rocky Mountains, in Europe as far as the mountain ranges adjoining the Mediterranean, and, in Asia, through the Himalayas. As might be expected in these circumstances, it is a very variable species—in fact a number of its regional forms have at various times been elevated to specific rank. This is not, however, justified and these are treated as subspecies in this account. As a native of this country, occurring in the hilly districts of Yorkshire, the Lake District, Wales, Scotland and Ireland, it has an additional claim on the affections of home Members.

It is interesting to read Roy Elliott's comments in his book *Alpine Gardening* that a pre-war radio census of the six best rock garden plants voted *S. oppositifolia* into third place. Since those days its stock has fallen somewhat, along with that of saxifrages in general; or perhaps it is just that it has been superseded by a large number of more recent contenders. It still remains, however, an outstandingly beautiful and easy alpine, once its not-too-complicated needs are understood. Perhaps some who buy it fail to appreciate sufficiently its dislike of a sun-baked position. *S. oppositifolia* relishes a cool but open site, where it will not get dried out and hence is well content with a northern or eastern-facing slope, but at the same time responding well to a good surface dressing of limestone chippings which act as a highly beneficial mulch. If one of the more robust forms such as *S. oppositifolia* 'Splendens' is selected and placed correctly on the rock garden, then this species can be rated an easy and rewarding plant worthy of the esteem in which it was formerly held.

When I read in various books of the necessity of working a fine top-dressing of sand and peat annually into the interstices of the mat of foliage, I am somewhat surprised that such elaborate care should be considered necessary. Though it is not specifically stated, I think that this can only have originated from the pan culture of the species where this routine has been found helpful in keeping the plants healthy. Although I have grown it in the alpine house, the frame and the open rock garden, I have found its ease of culture to be in inverse ratio to normal expectations based on experience with other choice alpines. My open-ground plants demand, and get, no attention whatsoever other than to prevent them being overlain by more rapidly-growing neighbours and, in spite of this, they flourish and give a wealth of bloom. It is even possible to detach pieces and replant them nearby to form new plants. All my plants, however, are placed on slopes—or where their mats can hang down over a rock, and it is my belief that this species, like *Aubrieta*, prefers

*Saxifraga manshuriensis* (p. 103)　　　　　*Photo: Roy Elliott, A.R.P.S*

*Saxifraga cymbalaria* ssp. *huetiana* (p. 114)

98                    *Photos: Winton Harding*

*Saxifraga petraea* (p. 115)

this method of planting. Pans in a frame do slightly less well, but still well enough to appear creditably on the show bench, whilst pans which I have in the past kept in the alpine house throughout the year, have proved difficult to keep in good condition—conditions were, I am sure, too warm for them in the summer.

It is worth mentioning that *S. oppositifolia* remains in flower—and an object of beauty—for a longer period than one might expect from a plant that begins blooming so early in the season, and this should be counted to its credit. Growing wild in Europe, there are five well-marked subspecies as well as the typical form. No doubt in Asia and North America there are other subspecies but none has, to my knowledge, reached this country.

In *Flora Europaea* Vol. 1 Professor Webb does not treat the variants of *S. oppositifolia* as subspecies but for convenience uses their original specific epithets, though this does not imply specific distinction. Here they are treated as subspecies.

*S. oppositifolia* ssp. *rudolphiana* (*S. rudolphiana* Hornsch. ex Koch) occurs in the Eastern Alps and is very compact and small in all its parts. Amongst other localities it is to be found on the Gross Glockner. It favours granitic formations and in cultivation is a less hearty grower than a number of the other subspecies, particularly those favouring calcareous formations. The flowers are rose-purple and quite handsome.

*S. oppositifolia* ssp. *blepharophylla* [*S. bleparophylla* Kerner ex Hayek] occurs in the Central Alps, particularly in Austria, and normally on granitic formations. The leaf has a very obtuse apex and bears long cilia on the apical half. The purple flowers are starry and less ornamental than those of some other subspecies. It is a high alpine, difficult to grow, and therefore of negligible horticultural importance when there are more beautiful forms which are considerably easier to grow.

*S. oppositifolia* ssp. *glandulifera* (*S. murithiana* Tiss.) occurs in south-west Europe and the Alps, normally on granitic formations. It is distinguished by the cilia of the calyx being gland-tipped. The flowers are purple and starry and the habit somewhat straggly. Again, it is of negligible horticultural importance.

*S. oppositifolia* ssp. *latina* [*S. latina* (N. Terracc) Hayek] is found only in the Apennines and is a very attractive form. Due to the presence of a larger than normal number of lime glands, the leaves are noticeably more encrusted than other subspecies. The large well-rounded flowers are a clear rose pink, with little if any suggestion of purple when in their prime. It favours calcareous formations and, being a good grower, is a very desirable plant.

*S. oppositifolia* ssp. *apennina* [*S. speciosa* (Dörfler & Hayek) Dörfler & Hayek] is found only in the Abruzzi mountains of Central Italy and is one of considerable merit. The habit is very compact. It is distinguished by the apical part of the leaf having a broad cartilaginous margin without cilia. Two excellent photographs, taken by Dr. R. Seligman, appeared in the *Bulletin*, Vol. 3, p. 102. He described the flowers as measuring in some cases "well over an inch (2.5 cms.) in diameter", and as "deep red". The petals are said to be frequently six in number and this certainly appeared to be the case in the plant photographed. Unfortunately, though I have visited the same place, it was later in the season and I missed this plant. I intend, however to return, for I am not at all sure that the true plant is any longer in cultivation in this country. Unless there are some unrevealed snags to its cultivation, it appears to be a particularly desirable form.

Whilst the above naturally-occurring forms have been given specific, sub-specific or varietal rank by botanists, there are a number of other forms that have been selected and propagated by horticulturists. The best cultivars are:

*S. oppositifolia* var. *alba* (illustration page 80). This white-flowered form is referred to rather disparagingly by Farrer and in the Royal Horticultural Society's *Dictionary of Gardening*. It is true that the individual flowers are only of medium size and rather narrow-petalled but, against this, they are a clear white and the plant is so floriferous that it is often covered almost completely by its blooms. It is suggested in the *R.H.S. Dictionary* that the starry form of the flower is derived from the less desirable subspecies *murithiana* and this, if correct, would explain a point that has always puzzled me. If it is not as strong-growing as some of the red or purple forms, it is certainly not lacking in vigour. I have had one plant in a pan for seven years and it has steadily improved with no special attention other than periodic repotting. A particularly good form of this variety was given the Award of Merit at the Society's Show on 9th April, 1963.

*S. oppositifolia* 'Splendens'. This form is sometimes referred to as *S. oppositifolia* var. *pyrenaica* 'Splendens', but, as there is no botanically accepted subspecies *pyrenaica*, this seems an unnecessary complication. It can be regarded as a particularly fine selected form from the Pyrenees, with large well-rounded flowers of a heather-purple colour and possessed of considerable vigour. It is the best grower of all the *oppositifolia* forms and has displaced most of them in commerce. It grows well on calcareous soils, is very floriferous and is a thoroughly good horticultural proposition.

*S. oppositifolia* 'W. A. Clarke'. This is a selected and extra fine colour form which has been grown for many years for its rich crimson flowers of medium size. It is a plant of neat habit but not by any means as robust a grower as *S. oppositifolia* 'Splendens'.

*S. oppositifolia* 'Wetterhorn'. This is another selected colour form with large rose-red flowers though not quite as fine as *S. oppositi-*

*folia* 'W. A. Clarke'. It is, however, a rather better grower than the latter.

A number of other selected forms have been named and offered by various nurserymen, but I do not consider them worth describing separately as they are only marginally distinct.

### S. RETUSA Gouan.

A particularly beautiful species, superficially resembling some of the compact and very prostrate forms of *S. oppositifolia*, but readily distinguishable by the erect corymbs of up to 5 small, but rich, rose-red flowers on 2-3 cm. stems. The purple stamens with orange anthers, which are longer than the petals, are prominent.

It is found at altitudes of over 2000 metres in the Pyrenees, Alps, and the Carpathian and Bulgarian mountains. Two subspecies are distinguished:—

*S. retusa* ssp. *retusa* (previously ssp. *baumgartenii*), has glandular-pubescent sepals and 1-3 flowers to a stem. It is usually found on non-calcareous soils throughout the range, except in the south-western Alps.

*S. retusa* ssp. *augustana*, has smooth sepals and 2-5 flowers, slightly larger than the previous, to a stem. It is usually found on calcareous soils in the South-Western Alps.

I have never seen *S. retusa* growing in the open in this country, though this might be more possible in the north. I have, however, admired most delightful pans of this species at the Shows on more than one occasion. It likes a gritty soil with plenty of humus, good drainage and ample moisture. It must on no account be baked or waterlogged. I have only recently started growing it myself and as it features in at least three nurserymen's lists it cannot be impossibly temperamental. *S. retusa* is a most dainty and charming plant well worth a little trouble. I feel that a thick top-dressing of granite chippings would probably help it considerably.

### S. BIFLORA All.

A species of the highest altitudes and the regions of the perpetually melting snows. It has the appearance of a small and rather straggly *S. oppositifolia* with more oval leaves, far less closely set along the stems. 2-5 flowers are borne in terminal inflorescences on the leafy shoots. The reddish-purple, or occasionally dull white, flowers have a certain attraction but in cultivation are rarely borne profusely enough to give much of a display.

This species occurs only in the Alps, where it is found in Austria, France, Germany, Switzerland and Italy growing by snow melts in much the same way as *Soldanella alpina*. As well as the normal form, there is also another one with considerably larger flowers, var. *kochii* Kittel [ssp. *macropetala* (Kerner) Rouy & Camus].

This species can be grown as a pan plant in this country using a carefully contrived compost and a considerable amount of care, though it will probably not survive for very many seasons. This plant would not seem to be good enough, compared with the other two species in this Section, to warrant the amount of care necessary to keep it going for long in cultivation. It must be rated a difficult plant.

---

## MICRANTHES SECTION

The Micranthes Section, formerly called the Boraphila Section, which includes upwards of 80 species, is not one that furnishes many plants of horticultural interest in this country. The species are distributed throughout the arctic, sub-arctic and temperate regions of North America, Northern Europe and Northern Asia, including Japan. In addition they crop up in the alpine and sub-alpine zones of the Himalayas and the adjoining mountain ranges of China and Tibet. In general, the more desirable species are extremely difficult to grow in this country whilst those which are relatively good-tempered are coarse plants of relatively little horticultural value. However, supplies of seed of a dozen or more species have been offered on numerous occasions in the Seed List, most of which have I believe been collected by our American members, so that some description of these species, at any rate, is justified. Furthermore, in the context of this review of the whole genus, it would be inappropriate to pass over such an extensive section, even though its contribution to the decoration of our gardens has, to date, been small. There is always the possibility that interesting species for the more specialist grower will, from time to time, be introduced, and on this count some sign-posts, indicating their place in the scheme of things, should be given. It is proposed therefore, to describe nineteen species briefly and in the case of the remainder merely to state which Sub-section they belong to.

Though I am currently attempting to grow a number of these species, my personal experience with them is, to date, distinctly limited. I have therefore gathered together as much information as is available from various more knowledgeable sources and, even if this is defective in places, I trust that these rather sketchy notes may in due course be of assistance to some of our Members.

There is no doubt that these species are more likely to flourish in the northern and western parts of these islands where two species, *S. nivalis* and *S. stellaris*, are in fact natives. Almost without exception they require shade, a moist atmosphere and a spongy soil with plenty of moisture in it. Some, such as *S. stellaris*, even grow well under bog conditions. A number will succeed quite well in moist woody conditions even in the south. Anyone expecting a spectacular display from these plants will be disappointed, for most have relatively small flowers, and foliage that is often coarse and uninteresting. The flowers are usually white or greenish-white with the occasional exception, such as *S. manshuriensis* with pale pink petals. The leaves are all radical, forming a basal rosette or rosettes and are often of a leathery texture and coarsely toothed. Some species are low plants with inflorescences only 5—10 cm. high, whilst others are over a metre in height with large spreading inflorescences containing very many small flowers. Most of the plants in this Section prefer an acid or neutral rather than a calcareous soil.

## Punctatae

### S. PUNCTATA

This widespread and variable species from North America and Asia has given its name to the Sub-section. It has a basal rosette of coarsely toothed kidney-shaped leaves and a flowering stem up to 50 cm. in height bearing a small number of not very striking white flowers. So far as I am aware, it is not now grown in this country.

It has a subspecies *arguta* which is sometimes listed in the Seed List as *S. arguta*. This differs only in its longer extruded stamens which are almost as long as the petals. In the Rocky Mountains it grows along the watercourses and is called 'The Brook Saxifrage'.

### S. MANSHURIENSIS

This species from North China, Manchuria and Korea was exhibited at the Autumn Show in 1967 by Mrs. Dryden and Mr. Piggin. There is a very good photograph (which is reproduced again on p. 97) except that, when photographed, the inflorescence had not yet opened out into the pretty umbel of soft pink flowers which later develops. The flowers have protruding stamens, very conspicuous red pistils and, unusually, are 6 to 8-partite. The radical kidney-shaped leaves are handsome, being dark green and glossy. According to the description then given in the *Bulletin*, it is not an easy plant to grow as it dislikes pot culture and resents disturbance. It requires plenty of water or the leaves will rapidly show their displeasure by going brown. The flower scape is ultimately about 35 cm. high. I am told by Mrs. Dryden that she now has it growing quite happily in a sheltered position out of doors.

Seed was offered in the Seed List in both 1968 and 1969.

#### S. MERTENSIANA

This desirable species featured in at least one nurseryman's catalogue in the 1930's, and has cropped up again in the Seed List in recent years. It is a native of the Rocky Mountains in America. The kidney-shaped, radical leaves are long-stalked and furnished with coarse rounded teeth. The flowering stem is 30—45 cm. in height and branches out into a very elegant panicle of white flowers with prominent red anthers. Like other species in this Section it requires a cool shady position. This is one of the Micranthes species that is most worth trying to grow for its decorative value. It has a variety *bulbifera* in which the flowers on the lower part of the stem are replaced by red bulbils.

Also belonging to this Sub-section are SS. *fusca, japonica, korshinskii, nelsoniana, nudicaulis, odontophylla, purpurascens, reniformis, sieversiana* and *spicata*.

### Davuricae

#### S. DAVURICA

This species from sub-arctic Siberia and North America has given its name to the Sub-section but does not appear to be in cultivation at the present time. Its leaves are wedge-shaped, narrowing gradually into the petioles and coarsely toothed at their extremities. The inflorescence of small white flowers is up to 18 cm. in height.

#### S. LYALLII

I have never seen this species growing, but I was recently interested to see some slides of it in its native habitat in the Rockies. In these it appeared to be quite an attractive, if not spectacular, small species. From a rosette of long wedge-shaped leaves closely resembling *S. davurica*, a branching inflorescence 15—30 cm. in height arises. The latter carries numerous attractive little white flowers. The leaves are said to be deceptively like those of *Primula minima*. It grows most happily in light shade. Seed is offered in the Seed List on occasions.

Also belonging to this Sub-section are SS. *astilbeoides, calycina, grandipetala* and *redowskiana*.

### Nivali-virginienses

#### S. NIVALIS

This is our native so-called 'Alpine Saxifrage', a rarity found on wet rocks in the mountainous parts of Great Britain. It has a wide range, occurring throughout arctic and sub-arctic Europe, Asia and America. The small greenish-white flowers are borne in a very characteristic, compact, rounded inflorescence at the top of a 5—15 cm. stem. The leathery ground-hugging leaves are more or less rounded but vary in shape. It is not a plant of any garden value nor one that takes kindly to normal garden conditions. Seed is, however, occasionally offered in the Seed List and, in spite of my comments, I am currently attempting to grow it!

## S. TENUIS

This has the appearance of a taller and more slender *S. nivalis* and, in fact, used to be considered a variety of the latter. It is more exclusively a plant of the far north and is found in Scandanavia, the Faroes, Iceland and Russia but not in this country. It calls for the same thoroughly moist but sharply drained conditions as *S. nivalis* but unfortunately such conditions are not to be found in most gardens. Seed is offered from time to time in the Seed List but its value or permanence as a garden plant is more than doubtful.

## S. RUFIDULA

This species from Alaska and the Pacific side of North America was grown before the last war in this country and was shown in 1936 at an Alpine Garden Show where it gained both Cultural and Botanical Certificates for its exhibitor, Capt. H. P. Leschallas. It is still occasionally offered in the Seed List, probably from wild-collected seed. It has short stiff stems about 6—8 cm. in height and pure white, starry flowers. It is worth trying to grow as it is said to be both attractive and fairly amenable, but it must be given moisture and shade.

## S. REFLEXA

This species also comes from Alaska and the arctic regions of North America. It has loose heads of small white flowers. So far as I know, it is not grown in this country but seed is occasionally offered in the Seed List.

## S. VIRGINIENSIS

This species, by contrast, comes from the Atlantic side of North America and is, given a reasonably moist shady place, somewhat easier to grow than those so far mentioned in this Sub-section. The ovate leaves, about 5 cm. long, are, however, rather coarse and uninteresting. The 30 cm. high branching inflorescence is composed of a very large number of small white flowers and has a certain decorative value.

## S. SACHALINENSIS

This species comes from eastern Siberia and Sakhalin Island. It is occasionally offered in the Seed List. It grows to about 30 cm. in height but is difficult and not particularly meritorious.

## S. HIERACIFOLIA

This is one of the few members of the Micranthes Section which is found in damp woodland in most of the mountainous European countries and, as such, is more likely to respond to cultivation in this country. Unfortunately with its inflorescence of disproportion-ately small greenish-white flowers on a 30 cm. high stem, allied to its coarse foliage, it has little appeal for adoption as a garden plant.

Also belonging to this Sub-section are *SS. aprica, careyana, californica, caroliniana, eriophora, fallax, marshallii, mexicana, oblongifolia, occidentalis, parvifolia, rhomboidea, tenessensis, texana, unalaschcensis* and *yezoensis.*

## Melanocentrae

All the members of this Sub-section come from Asia, mostly from the Himalayas and the neighbouring mountain ranges in China. None are, so far as I am aware, in cultivation in this country, nor is seed normally available.

Individual species will not be described but the following belong to this Sub-section. *SS. atrata, davidii, divaricata, dungbooii, gageana, lumpuensis, melaleuca, melanocentra, pallida, paludosa, parvula, pluviarum, pseudo-pallida* and *tilingiana.*

## Integrifoliae

### S. PENNSYLVANICA

This species from the Atlantic side of North America makes a large plant up to a metre in height and only suitable for the wild garden. The narrowly-oblong leathery leaves are up to 30 cm. in length and the relatively small greenish-white flowers are carried in profusion in a broad pyramidal inflorescence. It is a fairly easy species to grow with some shade. *S. forbesii* is very similar with flowers of a cleaner white.

Also belonging to this Sub-section are *SS. columbiana, fragosa, integrifolia, montanensis* and *oregana.*

## Stellares

### S. STELLARIS

This very widespread and variable species occurs over the whole of the sub-arctic region of the northern hemisphere and is also to be found in all the mountainous areas of Europe including these islands. It must be a familiar sight to all our Members who have gone plant hunting in the mountains, though they may not always have identified it. It occurs along every mountain rill over large areas where the spongy, boggy soil provides the conditions that it revels in. For this very reason it is very difficult to please in the garden where I have never seen it cultivated. Seed is, however, frequently offered in the Seed List for those who feel they might be able to simulate its native conditions.

It varies greatly but is commonly seen about 15 cm. in height, bearing an open panicle of up to a dozen very starry-white flowers— the petals have two yellow spots at the base and the anthers are pink. The leaves are rather characteristically elongated, wedge-shaped and toothed at the end. It is a dainty little plant but far from showy, for the flowers are relatively small.

S. CLUSII

This, in appearance, is like a hairy and magnified *S. stellaris* growing up to 30 cm. in height with leaves up to 15 cm. long. It is to be found, not uncommonly, in wet corners in the Cevennes, Pyrenees and Northern Spain and Portugal. This species would in all probability be extremely difficult to keep in cultivation and I know of no record of it being grown in this country. Our Members may well, however, encounter it on their mountain holidays. A viviparous variety, with many of the flowers replaced by leafy buds, occurs in Spain and Portugal. *S. clusii* may also be identified by the fact that the petals are unequal, 3 long and 2 short.

S. FERRUGINEA

For the last year or two, seed of this species from the Rocky Mountains and Southern Alaska has been offered in the Seed List and I have myself just sown it. I imagine though, that it may well prove very difficult to grow under our conditions.

S. MICRANTHIFOLIA (SYN. S. EROSA)

This large growing species attains a height of up to 80 cm. and bears a wide billowing inflorescence of greenish-white flowers. The lanceolate leaves are up to 20 cm. long. It comes from the Atlantic side of North America and has been grown in the past in this country without very great difficulty. It is, however, best placed in the woodland garden.

Also belonging to this Sub-section are SS. *birostris*, *bryophora*, *clavistaminea*, *foliolosa*, *leptarrhenifolia*, *leucanthemifolia*, and *newcombei*.

## Intermediae

S. TOLMIEI

This species from the Rocky Mountains and Southern Alaska superficially resembles a sedum more than a saxifrage. It forms spreading mats up to 60 cm. across with densely crowded, light green, succulent leaves about 1 cm. long and stems 4—8 cms. high, loosely few-flowered. The flowers are about 1 cm. in diameter and are white or light yellow. The plant is very floriferous and attractive in its native habitat. It normally grows in coarse, sandy or stony screes permeated beneath with icy cold waters from the melting snows, and in this zone it is often the only plant seen. Unfortunately it has a reputation for being almost ungrowable in the garden. Seed collected from the Cascade Mountains, Mt. Hood and Mt. Rainier has been offered in recent years in the Seed List but I cannot recollect that it has ever reached the show bench. It obviously needs plenty of moisture and very sharp drainage.

This is the only species in this Sub-section.

## Merkianae

S. MERKII

This species from Eastern Siberia and Northern Japan is not, so far as I am aware, in cultivation in this country. The plant from Japan is called var. *idsuroei* and was once thought to be a species in its own right. Seed of *S. merkii* was offered this year in the Seed List and, ever optimistic, I have sown it. In the wild it is said to make a very attractive low plant with neat foliage and panicles of fairly large, pure white flowers. If it can be grown, it should prove an attractive acquisition.

This is the only species in this Sub-section.

---

# DIPTERA SECTION

Though not one of the major Sections of the Genus from the point of view of either horticultural interest or the number of species involved, the Diptera Section contains several interesting plants, foremost amongst which are *S. fortunei* which has received both a First Class Certificate and an Award of Merit, and *S. stolonifera* (formerly *S. sarmentosa*) which is our old friend, familiarly known as 'Mother of Thousands'. The latter, together with its close ally *S. cuscutiformis*, is not everywhere hardy in the garden and, in consequence, is treated as a house or greenhouse plant. The other cultivated members of the Section can, however, all be grown outside except in the coldest parts of the country, but they should be given a sheltered corner as they are not amongst the hardiest of the saxifrages. They mostly flower very late in the season, in September or October, and the first severe frost is likely to disfigure both their attractive foliage and their flowers. This lack of complete hardiness and the lateness of flowering may be considered the drawbacks of the Section, but the latter can in some circumstances prove an asset when one is seeking subjects to prolong interest in the garden at this late date.

The greatest fascination of these plants lies in the unusual and striking formation of their individual flowers with the marked irregularity in length between the petals, one or two petals usually being three or four times the length of the remainder. The flowers are usually borne in profusion in elegant panicles standing well above the foliage. Some of the species spread by the formation of thread-like stolons or runners whilst others, which bear no runners, must be increased by division or seed. On this division of convenience I have separated the species in the descriptions which follow.

All the species in this Section come from China and Japan though *S. fortunei* does extend into Korea and the adjoining area of Manchuria. At the present time fifteen species are known as belonging to the Section, though only seven of these are in cultivation. There are no known hybrids but *S. fortunei* has a number of naturally occurring varieties, differentiated principally by their different leaf-forms.

Since the compilation of Engler and Irmscher's monograph on the genus *Saxifraga*, further work has been done on the new material brought back from South West China and the adjoining areas by Henry, Forrest and Wilson, and notes on the Section were published by Professor Bailey Balfour F.R.S. in the *Transactions of the Edinburgh Botanic Gardens* in 1916. I have, in this instance, followed his classification which, with one other addition, raises the number of species from nine to fifteen.

The cultivation of the species which can be grown out of doors all follows the same lines. I have already mentioned that frosts are liable to curtail their period of effectiveness and even cause death if they are not sited with some care. They are really plants of light woodland and it is quite useless to plant them in harsh, exposed conditions. They require a fairly sheltered corner with a little light overhead shade and a moist lime-free soil containing plenty of humus in the form of leaf-mould or peat. If you can manage to give them these conditions, you will find them plants of interest and personality. They can, of course, be grown in the protection of the alpine house or frame but they are not great lovers of pot culture and are really better planted out if this is possible.

### Non-stoloniferous species
S. FORTUNEI A.M. 1937. F.C.C. 1958

*S. fortunei* is the most commonly grown species in this Section cultivated out of doors in this country. Collected by Robert Fortune in Western China, but occurring also in Japan, Korea and Manchuria, it was sent home in 1863. Engler and Irmscher's monograph showed it as a variety of *S. cortusifolia* but happily it has now reverted to the original specific name commemorating its discoverer. Given the conditions already described as suiting the Section as a whole, that is a moist, lime-free soil, rich in humus, in a sheltered and partially shaded corner, it is not a difficult plant to grow. *S. fortunei* is particularly characterized by only coming into flower in October when one has almost given up hope of its forestalling the frosty hand of winter, which can so rapidly mar both its foliage and bloom. However, except in the coldest parts of the country, it usually manages to provide an effective display at a season when it has few competitors.

The pure white flowers, carried in graceful 40 cm. high panicles above the foliage, are produced in generous profusion. Each flower has five narrow, unspotted petals of which the bottom one, or sometimes two, are toothed and three or four times as long as their fellows. This inequality of the petals gives the flower a characteristic and orchid-like appearance.

The handsome, rounded and crenately-lobed, glossy leaves, which are about 8 cm. in diameter and leathery in texture, have 10 cm. long petioles embracing the stems. The bright green colour, with which they commence the season, becomes progressively tinged with bronze-red. The undersides of the leaves are a rich red throughout.

Three different varieties of this plant occur naturally in different parts of its geographical range, vars. *crassifolia*, *incisolobata* and *obtusolobata*, but the variations in leaf-form, on which this naming is based, are not of significant interest to the horticulturist.

The cultivars 'Rubrifolia' and 'Wada' are, however, valuable for the handsome red colour which suffuses their foliage throughout the season, 'Wada' having the slightly deeper colour. The former received the Award of Merit in September 1964.

### S. CORTUSIFOLIA

*S. cortusifolia* is closely related to the last species but flowers about three weeks earlier and is thus less liable to have its display curtailed by frost. Its flowers are smaller than *S. fortunei* and it is perhaps on this account that it has not attained the same popularity. It is a native of Japan, and was brought into cultivation by the firm of Veitch prior to 1874, from seed sent home by Maries. It does not occur in China, where it is replaced by several closely related species.

It can always be distinguished from *S. fortunei*, not only by the size of the flowers but also by the one or two long petals being completely untoothed. Also the upper smaller petals are broadly ovate, spotted and narrow abruptly to a claw. Like *S. fortunei*, it has fine glossy foliage of the same rounded and lobed outline, though this is somewhat variable.

There is a variety *rosea* with pink flowers, said to be an improvement on the type, but I do not know that this can still be found in this country.

### S. MADIDA

This species, also occurring in Japan, is very close to *S. cortusifolia* and, by some authors, is included within it. It differs mainly in the characters of the leaves, which are longer and more persistently hairy, and is neither better nor worse than *S. cortusifolia*. Whilst at one time offered in at least one nurseryman's catalogue, it is no longer available through commercial channels so far as I am aware.

### S. RUFESCENS

This species, which is a Chinese kinsman of the Japanese *S. cortusifolia*, was collected by Forrest in Yunnan in 1906 and introduced by Bees in 1908. It is shown in Engler and Irmscher's monograph under the name of *S. sinensis*. Its flowers, appearing in July, escape the risk of frost damage. *S. rufescens* can be distinguished by the densely hairy red flower shoots and by being the only Diptera species in which the red colouring also stains the petals. This species also seems to have passed out of the catalogues, though I have no doubt it still persists somewhere in cultivation. It would seem to be worth resurrecting.

Also amongst the non-stoloniferous species, but not yet introduced to cultivation are:

*S. aculeata* from South Western China (Yunnan)
*S. flabellifolia* from China
*S. henryi* from South Western China (Yunnan)
*S. imparilis* from South Western China (Yunnan)
*S. nipponica* from Japan
*S. sendaica* from Japan

There is no reason to suppose that any of the above species have as much horticultural merit as those already introduced.

## Stoloniferous species

### S. STOLONIFERA (SYN. S. SARMENTOSA)

Some readers may not recognize the old household favourite, Mother of Thousands, Roving Sailor, or Wandering Jew, under this name. For most of its career since it was introduced to this country before 1771, it has been known as *S. sarmentosa*, the name bestowed on it by Linnaeus, but unfortunately the rules of botanical priority have displaced this, and so we must fall into line. It comes from China and Japan.

Its main distinguishing features are its round prettily veined and marbled leaves and the slender red strawberry-like runners bearing a profusion of baby plantlets. The little white flowers have the two lower petals 3 or 4 times as long as the remaining three petals which are blotched with yellow and red. The stems and undersides of the leaves are red and hairy.

It is a very good window-sill plant and even prefers a position where it does not get too much sun. It will survive in poor starved conditions but repays a little care by producing more luxuriant and handsome foliage. With the multiplicity of more unusual and striking house plants that have been introduced in recent years, *S. stolonifera* now features but little in the florists' shops but, with its ease of propagation, it is likely to be passed from hand to hand for many years yet. This hardly endears it to the florist! If its claims are modest, it is certainly an extremely accommodating plant. It will even survive the winter, planted outside in a sheltered position, in the more favoured parts of the country, though it cannot be called a hardy plant in the full sense of the word.

S.s. var. *tricolor*. The leaves are beautifully variegated with red and white markings, but it is not as hardy as the parent species. It is a useful little plant for an edging or hanging basket in the greenhouse.

S. CUSCUTIFORMIS (illustration p. 80)

This is a smaller and less common relative of the better known *S. stolonifera* but it is nevertheless firmly established in cultivation. It has been called the 'Dodder-like Saxifrage', this being a translation of the Latin name denoting the resemblance of the delicate red runners to the stems of the Common Dodder (Cuscuta).

The rounded and bronzed leaves are beautifully netted with white veins and are somewhat fleshy. The flowering scape bears a panicle of numerous white flowers 2 to 3 on each branch which are not quite so markedly irregular as in most of the other Diptera species. Sometimes the 5 petals seem almost equal and at other times 2 or 3 of them are noticeably longer. It is thought to originate from China though it has not been found in the wild. This plant is hardy at Wisley but in the colder parts of the country it is probably best wintered in a frame or in the Alpine house. In any case the beauty of its leaves makes it an admirable plant to be seen at closer range on the bench of the alpine house.

S. VEITCHIANA

This species was collected in West Hupeh in 1900 by Wilson and introduced by the firm of Veitch in 1904. It is like a small form of *S. sarmentosa*, but with green leaves without the veining characteristic of the latter. It produces panicles of many white flowers on 10 cm. stems and these have the 1 or 2 long petals typical of the Diptera Section, the petals being blotched with yellow at their bases. It increases by forming little plants on thread-like runners. Unlike *S. stolonifera* and *S. cuscutiformis* it is fairly reliably hardy and in its unpretentious way, a pleasant little plant for an odd corner of the rock garden. It flowers in summer.

Also belonging to this stoloniferous group of species, but not yet introduced into cultivation, are:

*S. dumetorum* from South West China (Hupeh and Yunnan)

*S. geifolia* from South West China (Yunnan)

---

## MISCOPETALUM SECTION

The species in this Section, of which there are only four, are tufted shade-loving plants thriving in damp but not boggy conditions. *S. rotundifolia*, their most common representative, is a widespread plant in Southern Europe and all species are quite easy to grow, given suitable conditions. They are, however, rarely seen in gardens as their preference for damp, shady places, makes them unsuitable for open borders or the rock garden. It is unusual to see them advertised for sale or exhibited at the Shows. They flower in early summer in this country. Unlike the Robertsoniana Section, which they resemble, the flowering stems usually bear a few small leaves.

Brief notes are given below on the three species with any claim to horticultural interest.

## S. ROTUNDIFOLIA

This is the main species from which the others, rated at one time as varieties, diverge. It is a lover of damp, shady places, and is found in all the more elevated areas of Central and Southern Europe. It is represented in Russia by a plant which used to be called var. *coriifolia* but which is now accorded specific rank as *S. coriifolia* though it is of no horticultural importance.

*S. rotundifolia* has rounded, coarsely lobed leaves with long, slender petioles. The leaves are fleshy and hairy. The flowering stems are up to 40 cm. high and bear panicles of numerous, small, starry-white flowers, the petals usually being blotched at the base with yellow and conspicuously dotted in the upper half with red spots.

Although rarely seen in catalogues, the species is usually available in the Society's Seed List. It is readily raised from seed but, once raised, it must be given a damp, shady place, preferably in the wild or semi-wild garden. There it will prove quite a pretty acquisition. Its main flowering season is May and June.

A variety *heucherifolia* occurs in the Carpathians and Greece with slightly larger flowers and a rather variable leaf outline, said to be reminiscent of *Heuchera*. This has been grown in gardens in the past but has no very distinctive merit to commend it.

## S. TAYGETEA

This is very similar to *S. rotundifolia* except that the flower-stems do not exceed 20 cm. in height and the round basal leaves are lighter green and almost hairless. It is only found in Greece, Albania and possibly Southern Italy. The shorter inflorescence carries fewer but slightly larger flowers than *S. rotundifolia*.

I have recently raised a batch of plants from seed offered in the Seed List and found it to be a charming little species for a damp corner. Plants set out in sunny, dry positions promptly showed their displeasure by dying, but the remainder grew away well.

## S. CHRYSOSPLENIFOLIA

This is another close relative of *S. rotundifolia* occurring in the Balkans. Technically it differs in the absence of the cartilagenous border to the leaves, as well as in certain other details. It is neither to be found in catalogues nor the Society's Seed List at the present time and it would not seem to have anything to commend it above *S. rotundifolia* which it closely resembles.

# CYMBALARIA SECTION

Though this small Section contains only four species, there has been considerable confusion in the past about their naming and this still persists in some places and publications. All are annuals and two of the species are very similar. They are pretty little plants which seed themselves very freely in any damp position and make themselves particularly at home in the interstices of paving or in the plunge beds.

### S. CYMBALARIA (illustration p. 98 of *ssp. huetiana*)

This species from Roumania and South West Asia, reminiscent of the Common Toadflax, is usually encountered in the form of its sub-species *huetiana*, which has smaller flowers than the type and bluntly crenate or subentire leaves. The leaves of the type have 7 to 9 blunt triangular teeth.

*S. cymbalaria ssp. huetiana* is distinguishable from *S. sibthorpii*, which is described below, by the sepals being spreading and not deflexed until after the fruit has formed. It is exceptionally easy to grow and in the alpine house or frame it can rapidly become a nuisance, seeding into the plunge material or into other pots. In spite of these habits, it is a useful subject for odd corners and also for planting in troughs, which can often lack colour in summer. Seen at this height, the daintiness of the small leaves and flowers can be appreciated the better and, although only an annual, it looks quite in keeping with its more aristocratic neighbours.

### S. SIBTHORPII

Though Farrer states that *S. sibthorpii* usually does duty for *S. cymbalaria* under the latter's name, I have not found this to be the case in any of the stocks that I have examined. I have never grown *S. sibthorpii*, which is a native of Greece, but it is distinguishable from either of the subspecies of *S. cymbalaria* by the sepals being deflexed rather than spreading when the flowers are open. The flowers are also somewhat larger and a richer orange-yellow. It is said to be an easy little annual.

### S. HEDERACEA

This species is now very seldom seen in gardens. It has a weak, diffuse habit and, unlike the preceding two species, a host of very small white flowers. It occurs in Greece, Yugoslavia and Italy and has no garden merit.

### S. HEDERIFOLIA

Engler and Irmscher show this species as a native of Abyssinia, which seems a slightly unexpected vagary of plant distribution. It has not been introduced into cultivation.

# TRIDACTYLITES SECTION

The species in this Section are, with one exception, rather fragile looking annuals or biennials of very limited horticultural importance. *S. tridactylites*, a local native of this country, is an example.

### S. TRIDACTYLITES

The Rue-leaved Saxifrage is usually a very insignificant looking annual only about 3—8 cm. high, though occasionally found up to 20 cm. high. It branches from quite near the base. The basal leaves, which soon wither, are spathulate and entire, but the stem leaves are tri-lobed. The small white flowers are borne in diminutive panicles and, in general, the plant has no decorative or horticultural value.

It occurs throughout almost the whole of Europe and, in this country, grows locally on walls and on light sandy soils.

### S. ADSCENDENS (illustration p. 123)

This variable species, when seen in its best form, as in the illustration, is a pretty little thing, even though it is only a biennial. Botanists divide it into three subspecies, ssp. *adscendens, parnassica* and *blavii*. The distinctions are small but ssp. *blavii*, from the northwestern area of the Balkans, has the largest flowers which, as in the other subspecies, are pure white. *S. adscendens* not only occurs over the greater part of Europe but also extends to the Western Hemisphere.

It differs from *S. tridactylites* in being a biennial rather than an annual and in the basal leaves being 3 to 5-toothed ahd persisting until flowering time. Normally it also attains larger proportions than the latter, reaching 10—25 cms in height.

This species is rarely if ever grown in gardens, even though it can be quite decorative, as can be seen from the illustration. Whether it could produce this effect in the garden is an open question. Seed has been offered in the Seed List and this is a ready means of producing a stock. A well-drained position, or a pan in the alpine house, would be advisable.

### S. PETRAEA (illustration p. 98)

Like the foregoing, this species also is a biennial, occurring only in the Eastern Alps of Italy and Yugoslavia. It makes a diffusely branching mass of delicate stems about 10—12 cm. high, all covered with soft glandular hairs. The basal leaves are deeply divided into numerous toothed lobes but the stem leaves are less deeply dissected. The plant is very floriferous and this species has considerably more decorative value than its near relatives. In spite of Farrer's derogatory comments, it can be considered a worthwhile horticultural proposition, at any rate for the alpine house, which is where I have been successful with it.

It likes a calcareous soil and a shady position. In the open in this country, I fear that, with its soft hairy nature, it would not stand the winter well unless partially protected.

S. BERICA
This species used to be considered a variety of *S. petraea* but has recently been accorded specific rank. It occurs only in one locality near Vicenza in Italy. Apart from various other minor differences, it is regularly perennial, which makes it a potentially useful plant to grow: otherwise it has the same attributes as *S. petraea*. As far as I know, it is not at present in cultivation in this country. Like the former it favours calcareous soil and shade.

S. NUTTALLII
This North American species has never been brought into cultivation and, as far as I know, has no particular value.

S. PARADOXA
This species is shown by some authors in this Section, whereas others place it in a completely separate genus and call it *Zahlbruck-nera paradoxa*. In the *Flora Europaea* it is shown as a saxifrage. It is a fragile little plant with shining leaves and pale green flowers, found only in Austria and Yugoslavia. It has no horticultural value and is not to be confused with *S.* x *paradoxa* of the Euaizoonia Section.

## TRACHYPHYLLUM SECTION

All the species in this Section are mat-forming plants character-ized by narrow leaves with bristles along their edges. Though quite commonly encountered in the mountains of Europe and America, they are plants which do not take particularly well to cultivation nor produce much of an effect. It is most unusual for them to feature in nurserymen's catalogues. Nevertheless, for the sake of completeness, it is necessary to give a brief description of the Section's commonest members.

S. ASPERA
This species is frequently encountered in the Alps, Northern Apennines and Pyrenees, where it usually grows on non-calcareous soils between about 1500 and 2200 metres. It is a plant of the wetter areas in nature, but in the garden can be grown reasonably well in an open gritty soil, providing it is not allowed to dry out. It does not, however, usually flower very well in cultivation. The stiff, lanceolate, bristle-edged leaves stand out at a wide angle to the stems and there are many non-flowering lateral shoots, condensed almost into buds, which give the plant its dense habit. The cymes of 2—5 pale yellow, orange-spotted flowers are borne on 8—20 cm. high stems.

S. BRYOIDES (illustration p. 124)
This species, which used to be considered a variety of *S. aspera*, has the appearance of a more compressed version of the latter adapted for higher altitudes. It is, however, a more widely diffused species occuring from the Pyrenees to the Carpathians and Balkans at altitudes between 2000 and 4000 metres, almost always on non-

calcareous soils. I have grown it for a number of seasons without much difficulty but have only obtained the occasional flower in spite of its vegetative growth being quite healthy. A gritty, open soil with plenty of water suits it, but, unless my experience is exceptional, it is not a very worthwhile plant for cultivation. The foliage is similar to that of *S. aspera* but the flowering shoots are more compressed, so that the plant is even more mat-like. The flowers are always borne singly on 3—8 cm stems.

Seed is frequently offered in the Seed List, usually, I fancy, collected from the wild.

### S. BRONCHIALIS

This is usually regarded as a North American counterpart of the last two species, though in fact it is a plant of circumpolar distribution, occuring in Arctic and sub-Arctic Europe, Asia and North America. In Europe, however, it occurs only in Northern Russia. It is most commonly encountered as var. *austromontana* with flowering stems 6—15 cm. high or var. *rebunshirensis* with flowering stems only 4—6 cm high. The latter variety comes from Japan.

*S. bronchialis* is very close to the two foregoing species and of equally little horticultural merit. Seed of both varieties has been offered in the Seed List from time to time, usually collected from the wild, so that opportunities exist for the curious to try them.

### S. TRICUSPIDATA

This species comes from the extreme north of Canada, Alaska and Greenland. It is distinguished from the other members of the Section by having more wedge-shaped leaves which are 3-lobed at their extremities. The inflorescence, which is 5 to 15 cm. in height carries 1 to 3 creamy-white flowers, the petals of which are yellow in the lower half and lightly spotted with purple in the upper half.

*S. tricuspidata* is not difficult to grow in a moist, shady situation but it does not normally produce much of an effect in the garden. Seed is offered fairly frequently in the Seed List but I fancy it must come from Alaska.

*S. anadyrensis, S. cherlerioides, S. firma, S. nishidae*, and *S. spinulosa* are all species belonging to the *S. bronchialis* aggregate and only differing from it in minor points.

## TETRAMERIDIUM SECTION

This Section was formed by Engler to absorb a single species, *S. nana*, from the Chinese province of Kansu, which is distinguished by having flowers without petals and the other floral members in fours rather than the usual fives. It also has opposite leaves, yet plainly does not fit into the Porphyrion Section. Since then, however, another eight Himalayan species with opposite leaves have been discovered, some with the floral members in fours and some in fives. Consequently there is now a continuous range, some of which plainly belong to the Kabschia Section by any criterion. H. Smith, who has made a study of these species, feels that the only logical solution, with the increase in knowledge, is to absorb them all into the Kabschia Section and to do away with the Tetrameridium Section. However, these considerations are of little interest to the horticulturist, as none of these species has been established in cultivation and, judging from their elevated and barren environment in the Himalayas, it is doubtful whether any ever will be.

## XANTHIZOON SECTION

This Section contains only one species so that its characters are automatically also those of the Section.

S. AIZOIDES

This occurs throughout Europe, as far south as Central Italy and the Pyrenees, and also extends into North America and Asia. It is one of our own native saxifrages, found in the more northern and damper parts of these islands.

S. aizoides makes a loose mat of spreading branches, clothed with linear, light green, fleshy leaves, 1 to 2.5 cm. long, and with terminal cymes of bright yellow flowers spotted with red. It is also found less frequently with orange flowers, when it is called var. aurantia, or with dark red flowers, when it is called var. atrorubens.

In the wild it always grows under very wet but not stagnant conditions and is most often found alongside mountain rills or where the soil is sodden with spring water or melting snow. There it makes great mats with its roots down in the soil through which fresh water is continually passing. If we could all reproduce conditions such as these in our gardens, it would be a very easy plant to grow, and we should value it highly for the colourful and decorative species that it is. Alas, the garden that can provide such conditions is the exception and, under normal garden conditions, the species either soon fails or is only a very pale shadow of itself. It may be slightly more tractable in the north where the climate is more humid.

In the alpine house, where soil and watering are under complete control, it is a little easier to grow, and it is quite often seen on the show bench but never remotely resembling the luxuriant mass it becomes in the wild. A very shingly mixture is advised. Seed is offered in the Seed List in most years in the yellow, orange and red forms.

It has hybridized with S. umbrosa var. primuloides to give a well-known hybrid S. x primulaize which is considerably easier to grow than S. aizoides. This hybrid is obtainable in commerce and makes a pretty little plant resembling a small 'London Pride' with narrow leaves and short inflorescences of carmine or salmon-pink flowers.

S. aizoides has also hybridized in the past with S. mutata and S. caesia but the offspring, under the names of S. x hausmannii and

*S.* x *patens*, do not appear to be available any longer in commerce and, in any case, they never had much more than curiosity value as natural hybrids liable to be encountered in the wild.

## HIRCULUS SECTION

The species in this Section, which are all perennials, form mats of undivided, deciduous leaves from which arise flowering stems bearing varying numbers of yellow or orange flowers in late summer or autumn. The flowering stems are usually furnished with a few leaves. These plants mostly require a peaty soil and moist conditions throughout the growing season and even then are far from easy. One species, *S. hirculus*, is a rare native plant of these islands, usually found in boggy habitats.

The Section contains upwards of 170 species, most of which are found in the Himalayas and the mountains of Tibet, Burma and South-West China. A number, however, hail from the arctic and sub-arctic regions of North America, Europe and Asia. They make a very confusing Section with many species differing only in very minor botanical details. The greater number have never been introduced into cultivation, only having been collected as herbarium specimens. Most of the species introduced into this country have unfortunately been lost after a comparatively short time. This makes it difficult to say with any certainty how many species are now in cultivation—probably not more than a dozen at most, though I propose to give brief descriptions of 18 species. Fortunately four species, at any rate, seem now to be established in this country beyond any likelihood of disappearance, *SS. brunoniana, cardiophylla, flagellaris* and *strigosa*, and, if only on this account, the Section must have some space devoted to it. A further species, *S. pardanthina*, is still offered by one specialist nurseryman.

Because of their difficulty of cultivation, any members of this Section that become available present a challenge that is always likely to find some takers. Anyone attempting their culture must ensure conditions that will provide moisture throughout the growing season and in many cases some protection from excess damp is required during the winter. Probably the frame or open ground is more suitable than pot culture in the alpine house. Two of the species mentioned above throw out runners extensively and so require room for their runners to root.

A number of the species that have disappeared from cultivation were more or less monocarpic, such as *SS. candelabrum* and *umbellulata*. A bad seed-bearing season can eliminate such species completely unless seed or young stock is held in reserve. Seed should whenever possible be taken off all these Hirculus species as a safeguard against their being lost to cultivation.

Having grown only four members of the species personally, I have perforce had to search the available literature affecting them as well as consult a number of my colleagues to fill in the gaps. I am most grateful for the information I have been given.

Whilst the thought of the 150 or so species still awaiting introduction is a somewhat awesome one, I do not feel that they hold much promise for any but the most devoted specialist. He, however, should eagerly seize on the opportunity to get hold of and grow any that become available with the passage of time.

The Section is broken up into 9 Sub-Sections but of these only 5 are represented in the list of 18 species which are described separately below. These species are as follows.

| HIRCULOIDEAE | SS. *aristulata, cardiophylla, diversifolia, hirculus, palpebrata, pardanthina* and *turfosa.* |
| GEMMIPARAE | SS. *brachypoda* and *strigosa.* |
| SEDIFORMES | SS. *candelabrum, chrysantha, jacquemontana, pasumensis, signata* and *umbellulata.* |
| FLAGELLARES | SS. *brunoniana* and *flagellaris.* |
| HEMISPHAERICAE | S. *eschscholtzii.* |

## Hirculoideae

### S. ARISTULATA

This is a very dwarf and compact species from the Eastern Himalayas occurring freely in Sikkim and Nepal. Its narrow leaves are packed closely together in tight rosettes carrying handsome yellow flowers singly on 5 cm. stems.

### S. CARDIOPHYLLA (illustration p. 125)

This interesting species was originally discovered by the Abbé David in Western China in 1869. It is found in the provinces of Yunnan and Szechuan growing at about 3,500 metres. It was successfully grown and shown by Mrs. Dryden in 1966 and in answer to an appeal from the Editor of the *Bulletin* for further information the following was forthcoming.

Major General D. M. Murray-Lyon reported that he had obtained seed of it in 1953 from that wonderful grower of alpines, R. B. Cooke. He wrote "I find it easy in moist 'humus-y' soil where it flourishes alongside such plants as *Corydalis cashmiriana, Primula clarkei* and the Schizocodons; it makes a clump 10—12 cm. across and the same in height, flowering in September and October before dying down for the winter".

Mr. R. I. Smith of Westmorland considers it "by far the loveliest and easiest to grow of all Hirculus Section in cultivation today." He has a fine specimen measuring 20 to 30 cm. across, which produced about 30 flowering stems in August 1966 (about 4 years after he acquired it as a small pot plant.) Each flower stem carried

4 to 8 flowers of an orange-yellow colour. Seed is set in considerable quantity and it is growing in peat and sand in a shady position.

Seed has been offered twice in recent years in the Seed List and Members should attempt to grow this lovely saxifrage to ensure its continuity in cultivation. No doubt it will grow better in those parts of the country with a high rainfall. When well established it can in fact be divided and treated as a herbaceous plant.

## S. DIVERSIFOLIA

This is a very variable and widely distributed polymorphic species found in Yunnan, Szechuan and Upper Burma. It was first flowered in this country at Edinburgh and Kew in 1881. Irving in his book *Saxifrages*, written between the wars, considered it the most handsome member of the group and said that it had been cultivated regularly in gardens since its introduction. Farrer on the other hand had not much time for it except for the variety *foliata*. This variety, marked by handsome foliage with well marked veining of the ovate leaves, was shown to the R.H.S. Scientific Committee by Professor Bayley Balfour in 1912. At the present time I can find no proof that either the species or its variety are in cultivation in this country though I may be mistaken.

It grows to a height of about 25 cm. but the large oval leaves are in most forms not particularly attractive and rather tend to overpower the effect of the 1 to 2 cm. diameter golden-yellow flowers.

Its seed has been offered in the Seed List in recent years though I cannot vouch for its trueness to name. It is reputed, according to some authorities, to be less difficult to grow than many of the Hirculus Section.

## S. HIRCULUS

This species is spread widely over the Northern Hemisphere, particularly in the more northerly latitudes and is a rare plant in the wetter parts of these islands. The form which occurs in the Himalayas has, on account of minor differences, recently been described as a separate species under the name of *S. montana*.

From its tufted root it throws up flower stems about 15 cm. in height, clothed with narrow leaves and bearing solitary yellow flowers about 2 cm. in diameter. The petals are covered with orange spots in their lower halves. There is also a very handsome variety *major* from Central Europe which grows up to 30 cm. high and has a flower nearly 3 cm. in diameter.

*S. hirculus* is a plant of boggy marshy areas and wet grassy slopes and, on this account, not easy to please in the garden, though handsome when it can be given the right conditions. Probably a mixture of sand and peat, which is not allowed to dry out, offers the best chance of success. I have never seen it shown, nor growing in gardens, though there may well be gardens where it flourishes.

Certainly seed has been offered in the Seed List on at least three recent occasions though two of these offerings were marked as having been collected in Alaska.

S. PALPEBRATA

The Royal Horticultural Society's *Dictionary of Gardening* describes this species from Sikkim as having hairy leaves and stems about 5 to 7.5 cm. high bearing medium sized golden-yellow flowers. I suspect it is no longer in cultivation in this country.

S. PARDANTHINA

This species from the South-western provinces of China was first described in 1935 and so is of relatively recent introduction. It grows to a height of up to 30 cm. and has flowers of deep orange marked inside with crimson. It is currently offered by Reginald Kaye from whom I have procured it and so far it is growing well. He tells me that it prefers an acid peaty soil. By the time these words appear in print I hope that it will have flowered successfully. Like many others in this section it dies down completely in the winter and is then difficult to see.

S. TURFOSA

This is another spreading species from Yunnan which was in cultivation at the time that Irving wrote his book on saxifrages. The R.H.S. *Dictionary of Gardening* describes it as growing to a height of about 15 cm. with panicles of several yellow flowers and the habit of forming stolons from the base of the stems. Since the war I can find no record of it having been grown in this country so that it may be lost again now.

## Gemmiparae

S. STRIGOSA

This species is referred to first in this Sub-Section as it is quite incontrovertibly in cultivation at the present time. In the September 1968 *Bulletin*, Rhinanthus quotes D. B. Lowe to the effect that:—
"The flowers are buttercup yellow, about 1 cm. across, with tiny scarlet tips to the stamens and the hairs are red. This valiant little plant begins flowering in July and sustains a mass of buttercups until mid-October. It is superbly weather-proof and well-behaved. The petal formation, when studied closely, is delicate and curious."
Since that date, however, Mr. Lowe tells me that he has lost the original plant in a bad winter when it was frozen solid for six weeks. It survived however from seedlings which subsequently appeared. He now feels that this initial success may have been rather deceptively easy. It is probable that this plant is best treated as a fairly short-lived species under our conditions and that for safety, it should be raised frequently from seed. Seed which I have obtained from Mr. Drake has germinated rapidly and plentifully.

Also in June 1968 this species was shown by Mrs. Dryden in

*Saxifraga adscendens* (p. 115)
E. Alps

Photo: M. G. Hodgman

*Saxifraga bryoides* (p. 116)
(Dolomites)

*Photo: M. G. Hodgman*

*Saxifraga cardiophylla* (p. 120)    *Photo: Roy Elliott, A.R.P.S*

Saxifraga brunoniana (p. 129)

126                                    Photos: R. Elliott

Saxifraga flagellaris (p. 128)

London when I was much impressed by it. Since then she too has found it difficult to please. One feels from Mr. Drake's experience that it is far more at home under Scottish conditions than in the South. It is a native plant of Nepal, Sikkim and Yunnan.

S. BRACHYPODA

This is another species from Nepal, Sikkim and Yunnan which has in the past been in cultivation qut which has now equally certainly disappeared. It has narrow leaves and solitary golden-yellow flowers on 7—10 cm. stems.

## Sediformes

S. CANDELABRUM

This is a monocarpic species from Yunnan which has also now been lost to cultivation. It makes an interesting rosette bearing starry yellow flowers on a 15—20 cm. stem. The yellow petals are narrow and the green sepals appear prominently between the petals.

S. CHRYSANTHA

A passing mention is made of this species as, unlike the foregoing, it is a plant of the Rocky Mountains in America. As far as I know it has never, so far, been cultivated with any success in this country, though it is very noticeable to any visitors to its native haunts in Colorado and elsewhere. It is a plant of moist soils and bears 1 to 2 large golden-yellow flowers on very short stems 2—4 cm. high.

S. JACQUEMONTIANA

This charming species comes from Sikkim and Kashmir but has never to date been successfully grown in this country. There is, however, a very lovely photograph of it in the *Bulletin* (Vol. 30, p. 205). In the text Polunin describes it as: "A charming little saxifrage growing beside his tent in the Karakoram and forming a flat cushion on the soil. It has glandular rosettes about ¾ cm. across, closely pressed together and from the centre of each grows a bright orange-yellow flower, stalkless and spreading over the rosette from which it springs. On the inside of the petals, faint scarlet spotting shows".

S. PASUMENSIS

This species was, rather surprisingly, given the Award of Merit, when shown on 3rd May, 1949 by Mr. R. S. Masterton. Though first collected by Kingdon Ward on his 1924—1925 expedition to the Eastern Himalayas and Tibet it was actually introduced into cultivation by Major George Sherriff in about 1944. The description, when it was given the Award, ran as follows:—

"*S. pasumensis* consists at first of a single rosette up to 7.5 cm. across and resembles a miniature *S. florulenta*. The rosette is composed of numerous, closely imbricated, spatulate leaves which are greyish-green in colour and edged with glandular hairs. From the centre of the rosette arises a densely-branched flowering stem bearing a few small leaves and numerous large showy buttercup-yellow flowers with pointed petals. After flowering, the rosette dies but meanwhile a number of small lateral rosettes have been formed at ground level and these will flower the following year. The plant needs careful watering after flowering or it may rot away."

Unfortunately this plant seems now to have disappeared from cultivation after being grown by a number of people for some years after its introduction. It was last grown at Edinburgh Botanic Gardens, I am told, in 1958. There it tended to be monocarpic.

### S. SIGNATA

This species was first collected by Forrest in Yunnan in 1910 but the introduction into cultivation seems to stem from seed sent back from his 1930 expedition. It has also been found in western Szechuan as well as north-western Yunnan.

The dense basal rosette is from 3 to 5 cm. across and is composed of a number of fleshy spatulate leaves from which arises a leafy flower stem 8—18 cm. high carrying a spreading panicle of rather striking flowers. The petals are pale yellow in the apical half and greenish yellow with bands and blotches of red in the basal half. It is a dainty and desirable species but markedly monocarpic and, due no doubt to this, it seems now to have disappeared from cultivation.

### S. UMBELLULATA

This species from south-eastern Tibet and Sikkim was introduced by Major George Sherriff in 1938. It is close to *S. pasumensis* and the two were often confused one with the other. It now seems to have disappeared from cultivation.

## Flagellares

### S. FLAGELLARIS (illustration p. 126)

The species is particularly marked by the profusion of red strawberry-like runners which it sends out from near the base of the stems after flowering. Roy Elliott gave a most excellent description and photograph of this species in the *Bulletin* (Vol. 34, p. 162), from which I quote the following:—

"*Saxifraga flagellaris*—though not easy to grow—possesses remarkable beauty in its clear butter-yellow compact flower spikes. It really is a beauty; the difficulty seems to lie in keeping it through the winter, for it combines a love of summer dampness with a strong resentment of winter wet—thus making it essentially a plant for the controlled conditions of the alpine house. Each runner carries a tiny rosette, which quickly develops and roots with extraordinary ease into damp chippings (at any rate in the writer's alpine house plunge material). One must wait until the rosette has developed and rooted well before breaking the cord which links it to the mother plant.

*S. flagellaris*, though quite a rarity in cultivation, has a remarkably wide distribution through the northern and arctic regions of Europe and Asia, the Himalayas and Canada and even through the Rocky Mountains as far south as Arizona."

Of recent years the European plant has been separated as *S. platysepala* and the Himalayan plant as *S. komarovii* but these fine distinctions need not affect gardeners.

There are 3—4 relatively large bright yellow flowers to each lax cyme and these cymes are borne in great profusion which makes it a most decorative plant.

### S. BRUNONIANA (illustration p. 126)

This species throws out an even more amazing mass of glistening red runners than the last and again Roy Elliott gave a most excellent description and photograph of it in the *Bulletin* (Vol. 34, p. 162). To quote: "The ugly sister—perhaps that is being unfair—*S. brunoniana*, is found only in the Himalayas and was brought into cultivation in 1906. The flowers are produced in July in rather loose panicles on four or six inch stems (10—15 cm.) and are a rather washy yellow. It is reasonably hardy and even if it gets knocked about in the winter, the small rosettes will soon grow again in the spring-time. Its beauty lies more in the fascinating web of thread-like crimson runners than in the flowers but those who like 'interesting plants' will derive pleasure from both."

It is worth mentioning that the leaves which make up the rosettes are noticeably bristle-margined. Roy Elliott's remarks on its hardiness may need to be treated with caution in some parts of the country where it is inevitably killed if left out of doors in the winter, though it has apparently survived in his garden without protection since 1964. (It is still alive, 1970. Ed.)

### Hemisphaericae

#### S. ESCHSCHOLTZII

This species from arctic and sub-arctic America does not seem to have been established in this country, though seed from Alaska has been offered in the Seed List on a number of occasions in recent years. It is a compact plant making wide mats in its natural habitat. The flowers however are inconspicuous with tiny white or yellowish petals according to Mrs. L. Strutz who has collected the seed. She explains that it is a plant for the collector rather than for 'show'.

# Saxifrages Index

P *in heavy type denotes illustrations*